CAMBRIDGE COUNTY GEOGRAPHIES

SCOTLAND

General Editor: W. Murison, M.A.

T0352331

KINCARDINESHIRE

Cambridge County Geographies

KINCARDINESHIRE

by the late

GEORGE H. KINNEAR, F.E.I.S.

Headmaster, Glenbervie Public School

Author of *Glenbervie, the Fatherland of Burns*

With Maps, Diagrams, and Illustrations

CAMBRIDGE

AT THE UNIVERSITY PRESS

1921

CAMBRIDGE UNIVERSITY PRESS
Cambridge, New York, Melbourne, Madrid, Cape Town,
Singapore, São Paulo, Delhi, Mexico City

Cambridge University Press
The Edinburgh Building, Cambridge CB2 8RU, UK

Published in the United States of America by Cambridge University Press, New York

www.cambridge.org
Information on this title: www.cambridge.org/9781107649705

First published 1921
First paperback edition 2013

A catalogue record for this publication is available from the British Library

ISBN 978-1-107-64970-5 Paperback

PREFATORY NOTE

MR KINNEAR's death when he had written the text of this volume, but had not finally revised it, left the work to be completed by the general editor. Fortunately, the changes which Mr Kinnear intended to make were clearly marked ; and these have been closely followed. As it has been impossible to find an accurate list of those who, by advice or otherwise, assisted Mr Kinnear, will all who did so kindly accept this general acknowledgment of their much-appreciated help ?

The general editor is deeply indebted to his friend, Mr J. B. Philip, himself a son of the Mearns, who has given unstintedly of his full knowledge of the county and has rendered invaluable service in the reading of the proofs. In addition, Mr Philip generously presented a number of his own photographs for use in illustration.

W. MURISON

November 1920

CONTENTS

ILLUSTRATIONS

ILLUSTRATIONS xi

The illustrations on pp. 3, 21, 25, 26, 34, 35, 41, 58, 76, 115 are reproduced from photographs by Mr J. B. Philip; those on pp. 9, 15, 54 from photographs by W. Holmes & Co.; those on pp. 11, 97 from photographs by Mr Archibald Taylor; those on pp. 17, 33, 64, 66, 86 from photographs by Valentine & Sons, Ltd.; those on pp. 31, 37, 38, 62, 68, 75, 89, 90, 92, 93, 104, 106, 112, 117 from photographs by Mr A. Ross; those on pp. 39, 72 from drawings by Mr J. Reid; that on p. 67 from a photograph by Messrs Oliver & Boyd; those on pp. 70, 71, 73, 87 are reproduced by permission of the Society of Antiquaries, Scotland; that on p. 77 by permission of the Committee of Management, Free Public Library and Museum, Paisley; that on p. 79 by permission of *The Aberdeen Daily Journal;* that on p. 81 is from a photograph by Mr F. C. Inglis; that on p. 105 is reproduced, by permission of the University Court of the University of Aberdeen, from a photograph by Messrs T. & R. Annan & Sons; those on pp. 109, 111, by arrangement with the same firm; that on p. 108 from a photograph supplied by Mrs Kinnear, and that on p. 114 from a photograph by Mr W. J. Johnston.

1. County and Shire. The Origin of Kincardine and Mearns

The word *shire* is of Old English origin and meant office, charge, administration. The Norman Conquest introduced the word *county*—through French from the Latin *comitatus*, which in mediaeval documents designates the shire. *County* is the district ruled by a count, the king's *comes*, the equivalent of the older English term *earl*. This system of local administration entered Scotland as part of the Anglo-Norman influence that strongly affected our country after the year 1100.

The number of counties has not always been the same, nor have the boundaries always been as they are now. Geographically Kincardineshire and Forfarshire are one ; and in a very old account of the district it is stated that " Angus and Mearns were united and both called by the same name." The official who represented the King's authority was the Shire-reeve or Sheriff, but sheriffdoms were modified in number and area from time to time as was found convenient. Early in the fourteenth century there were at least twenty-five counties in Scotland, at the present time there are thirty-three.

The county was manifestly named from Kincardine in Fordoun parish, once a town with a royal residence. The name *Kincardine* is taken to mean " the end of the

A

high land," *i.e.* where the Grampians terminate. *Kincardine* occurs frequently as a place name along the east of Scotland from Ross-shire to Fife ; and, with the exception of Kincardine-on-Forth, it regularly designates a place at the end or the side of hills.

The county is often spoken of as " The Mearns," although this is not strictly accurate. The Mearns constitutes the district of the county south of the Grampians. The Howe of the Mearns is really a continuation of the great valley of Strathmore. Like " The Merse " and " The Lothians," we say " The Mearns," not " Mearns " alone. The etymology of *Mearns* is disputed. A tradition is that Kenneth II., in the ninth century, divided this region into two, bestowing them on his brothers Æneas and Mernas, whence they were called respectively Angus and Mearns.

2. General Characteristics

Kincardine, like its southern neighbour Forfarshire, of which it is indeed but a continuation, exhibits a good epitome of typical Scottish scenery. The two counties present pretty much the same physical appearance. Each, in a restricted though real sense, may be termed

> " Land of brown heath and shaggy wood,
>   Land of the mountain and the flood."

They both show a fertile tract of level or gently undulating land along the coast, dotted here and there with green plantation, stately mansion, or comfortable-looking homestead. In both counties, also, the interior

Coast Scenery at Muchalls

is well sheltered from the biting east winds that sweep in from the sea by a range of hills running parallel to the coast—the Sidlaw Hills in Forfarshire, and the Garvock Heights in the Mearns. Similarly, on the north side of each, the Grampians rising in majestic grandeur form a wall of protection from the cold northern blasts. In both, we thus have favourable conditions for the production of fertile soil through the disintegration of the rocks and stones on the hillsides, and through the age-long washing down by rain and flood of new soil from the " everlasting hills " into the valleys below. For the sportsman the hills and moors of Kincardineshire provide grouse and other game. The parish of Strachan contains the one deer forest in the county—the most easterly deer forest in Scotland.

The highly picturesque scenery along the coast of Kincardineshire is a never-ending delight to the artist and to other lovers of nature ; while in its diversified flora, its rock structures, its antiquities, the county offers ample material to botanist, geologist, and archæologist.

Kincardineshire has long been connected with the fishing industry, but the introduction of steam trawlers and drifters has, to a large extent, displaced the line fishing which was successfully pursued from the many villages and creeks along the coast. Manufactures can hardly be said to exist in the county. With the decay of handloom weaving, the manufacture of linens and woollens was transferred to the larger centres in the south and north—Dundee and Aberdeen. On the outskirts of the Forfarshire linen-manufacturing area, and connected with Dundee as the principal market-centre, there are, however, flourishing spinning mills at Bervie, Gourdon, and Johnshaven.

The county is intersected by the main line of the Caledonian Railway, which, at least for part of its distance, runs near to the route of the old roads, a fact which indicates the limitations imposed by nature both on the ancient makers of roads and the modern makers of railways. The position of many of the towns and villages is along this natural route. A similar explanation applies to the position of the towns and villages along the coast and in the Dee valley. With the exception, however, of Stonehaven and Banchory, their size has not yet greatly increased under the influence of railways, as in other parts of the country.

Although at one time possessed of a royal residence, Kincardineshire cannot be called a county of much national importance. On the other hand, few districts have afforded such an interesting field for the study of local history or research into manners and customs of the past. The county has a world-wide reputation as a sanatorium centre, while Stonehaven with its bracing air, its woods, walks, and scenes of beauty, and its unique opportunities for healthy recreation and enjoyment, attracts visitors from all parts of Britain during the summer months.

3. Size. Shape. Boundaries

The county is not a large one, but its area is compact and well defined. It ranks as twenty-first of the Scottish counties in extent, twenty-fourth in population, and twentieth in point of rental. From southwest to north-east it is 32 miles in length, and 24 miles, where widest, from south to north. It lies between latitude 56° 46′ and 57° 9′ N. and between longitude 2° 4′ and 2° 44′ W. The area of the county is 248,195 acres, or approximately 388 square miles. It is only one-eleventh of the area of Inverness, the largest county in Scotland, but it is almost eight times larger than Clackmannan, the smallest.

Wedged in between two bigger neighbours, Aberdeenshire and Forfarshire, Kincardineshire in shape resembles a right-angled triangle, the right angle being at Mount Battock in the west, while the two sides containing it are lines which run, roughly speaking, along the course of the river Dee to Aberdeen, and along the west side of the county towards the mouth of the North Esk. The other side, formed by the coast-line from near Montrose to Aberdeen, has a distance of about 35 miles. The whole outline measures about 100 miles.

The watershed of the Dee on the north, and the watershed of the North Esk on the west, practically mark out the county limits. The area lying between these two rivers and the sea comprehends a district the general slope of which is to the south-east. A picturesque background to the district is formed by the Grampian heights, varying in elevation from 500 to 2500 feet. From the summits of this natural barrier of hills, covered with heath and moss, there is a regular succession of green hills and cultivated slopes down to the Howe of the Mearns with its flat or undulating fields. The eastern boundary—the North Sea shore—runs at first to Bervie, a distance of 10 miles, in a north-easterly direction ; for the next 10 miles, to Stonehaven, it curves to the north ; and for the remaining 15 miles, to the Dee, it again takes a north-easterly direction. Along the north side from Aberdeen the Dee forms the dividing line as far as Crathes, a distance of 14 miles ; after which the county boundary sweeps round the north side of Banchory over the Hill of Fare, touching here the southern Aberdeenshire parishes of Echt, Midmar, Kincardine-O'Neil, and Birse. The western boundary from Mount Battock to near Montrose is formed by the North Esk, and the Forfarshire parishes of Lochlee, Edzell, Stracathro, Logie Pert, and Montrose.

Before 1891 the parishes of Banchory-Ternan, Drum-oak, and Banchory-Devenick were partly in Kincardine-shire, partly in Aberdeenshire. In that year Banchory-Ternan was all included in Kincardineshire, and Drumoak in Aberdeenshire. The designation Banchory-Devenick was now restricted to the Kincardineshire portion, while the rest was added to the Aberdeenshire parish of Peterculter. At the same time the parish of Edzell, which had been partly in Forfarshire, partly in Kincardineshire, lost its Kincardineshire portion, which was transferred to the parish of Fettercairn.

4. Surface. General Features. Soil

A bird's-eye view would show the county to be divided into two parts of nearly equal size, but of totally different aspects. As a whole it is very diversified, embracing districts that are entirely of a Highland type of scenery, while in the south and east the lowland and the maritime type predominate. In appearance it resembles Forfarshire, which, however, excels in extent and boldness of mountain summits.

The two main divisions of the county may be further regarded as being naturally sub-divided into four longitudinal and parallel districts: the Maritime, the Howe, the Grampian, and the Deeside. The Grampian district stretches through the whole breadth of the county from west to east, and is on an average from 16 to 18 miles in length and from 6 to 8 miles across from south to north. It is naturally rugged, sterile, and dreary. Its total area may be reckoned at 120 square miles, mostly moor and heather. From Mount Battock (2555 ft.) in the north-west, the highest peak in the district, the Grampians gradually descend and are popularly regarded as terminating in the low heights (about 200 ft.) near the Bay of Nigg. Of the other Grampian peaks the most prominent are Clochnaben (1944 ft.), with its granite knob 100 ft. high, a well-known landmark from the sea; Kerloch (1747 ft.), and Cairnmonearn (1245 ft.). The Hill of Fare on the north side of the Dee reaches a height of 1429 ft., while on the opposite side of the river, Scolty Hill (982 ft.), with its monument to General

View of the Feugh Valley from Scolty
(Clochnaben in right background)

Burnett, is a notable landmark, from which, as from the top of most of the hills, splendid views are obtainable of the scenery of Aberdeenshire on the north, and of the Howe on the south. .

The Deeside district extends westward from the mouth of the Dee along the southern banks of that river for 23 miles, and has an area of about 54 square miles. There is also a portion of this district on the north bank of the Dee above Banchory, extending to about 26 square miles. Although, as regards agriculture, Deeside is a comparatively poor region, yet there is in it a greater proportion of surface under timber than in any other part of the county. This gives it a very pleasing aspect, embellished as it is by the waters of the " Silvery Dee," flowing along through the level haughs and meadows that lie between the encircling slopes on both sides of the valley.

The Howe of the Mearns district, about 16 miles long, 5 miles broad, and having an area of about 50 square miles, forms the eastern boundary of the Vale of Strathmore ; but, in comparison with it, the Howe is very flat and bare, especially towards the eastern extremities. The soil here has a characteristically red appearance, due to the underlying clay, popularly known as " Mearns Keel."

On the lower slopes of the Grampians and overlooking the splendid panorama of scenery in the Howe are several well-known peaks. From west to east these are the Cairn (1488 ft.) ; Whitelaws (1664 ft.) ; Houndhillock (1698 ft.) ; Balnakettle Hill (1000 ft.) ; Garrol (1035 ft.) ; Arnbarrow (1060 ft.) ; Strathfinella (1358 ft.) ; Tipperty

(1042 ft.) ; Herscha (725 ft.) ; and Knock Hill (717 ft.). On the north side of Strathfinella lies one of the most romantic and picturesque spots in Scotland, the famous Glen of Drumtochty, leading by the " Clatterin Brigs " to Fettercairn and the Burn.

Between the Howe and the coast a lower range of

The Ford, Drumtochty Glen

hills, known as the Garvock Heights, cultivated almost to their summits, runs in a south-easterly direction through the parishes of Arbuthnott, Garvock, and St Cyrus. Their elevation ranges from 500 to 900 ft. Johnston Tower (915 ft.), immediately above Laurence-kirk, is a conspicuous object from all points of the compass. The view from this point is magnificent, comprising hill and dale, stream and sea, fertile fields and cosy homesteads. The descent from the Garvock

Heights to both the Howe and the sea is gradual, the slopes on each side being here and there dotted with patches of wood or grassy moorland which give freshness and colour to a somewhat bare and monotonous district. East of the Bervie valley a lower range, really a spur of the Garvock Heights, starts at Carmont Hill (710 ft.), on the south side of the Carron Water, and is continued, with a slight descent, over Bruxie Hill (700 ft.) to Bervie Brow (451 ft.), overlooking Bervie Bay.

When the character of the soils of the county is considered, regard must be had to the nature and structure of the underlying geological formation, since the soils are indebted to disintegrated rocks for their mineral constituents. The quality of a soil, moreover, depends largely on the upper formation of the neighbouring heights, the decomposed portions of which are washed down by rain and flood to the lower grounds and there incorporated in the soil.

The best farming district in the county is undoubtedly along the coast, especially between St Cyrus and Bervie. The soil here is a deep black loam. Most of it is free, parts of it tenacious, but none of a stiff, clayish nature. The farms in the St Cyrus district are undoubtedly the most fertile in the county. The liberal application of lime from the lime-kilns at St Cyrus in the beginning of the last century was a valuable factor in improving the soil ; and this, combined with intelligent and up-to-date methods of cultivation, is reflected in the general excellence of the crops in this district. In striking contrast are the thin and cold soils of the Garvock region. Owing to elevation, exposure, and absence of thorough drainage in many parts, the soil cannot be called a very kindly one, although during the last half century much has been done to improve the appearance of this somewhat bare district.

The soils in the Howe of the Mearns are rather variable, ranging from deep brown soils resting on the boulder clay, to a light gravelly moorish soil stretching right up to the middle of the Howe, through part of the parishes of Marykirk, Laurencekirk, Fettercairn, and Fordoun. In the neighbourhood of Laurencekirk the soil is a stiff clay, though on the whole a good cropping soil. The Grampian district, the largest in the county, is naturally a poor farming district, the soil being neither deep nor productive. Still, on the clay-slate formation, we find along the southern spurs of the Grampians a soil remarkably well-adapted for the growth of timber plantations, which are here very numerous.

The soil on Deeside is of a thin, gravelly nature, being formed from decomposed schists and granite rocks, with a small proportion of moss or decayed vegetation. Though not suitable for the production of heavy crops such as wheat and barley, the soil of Deeside is remarkably suitable for the culture of small fruits (strawberries and raspberries), and especially for the growth of timber. Between Deeside and the coast, in the district around Muchalls, much of the soil is mossy, while between Stonehaven and Aberdeen a strip of land, bordering the sea and extending a few miles inland, is of the same nature. In the neighbourhood of Aberdeen, part of the land is laid out for the cultivation of potatoes and vegetables, a ready market for which is got there.

5. Rivers and Lakes

With the exception of the Dee and the. North Esk, which belong only in part to Kincardineshire, the rivers of the county are comparatively small. But from its diversified and unequal surface, and from the fact that the land slopes in many directions, the streams are numerous, and every part of the county is well-watered. By reclamation, drainage, and improved cultivation, and the consequent disappearance of water-logged haughs adjacent to rivers, the streams are smaller than formerly, but admirably adapted for the natural and the artificial drainage of their districts. The course of the streams flowing from the Grampians to the North Sea shows the general slope of the county to be towards the south-east. The northern district is drained by numerous tributaries discharging into the Dee, and the western by the North Esk and its feeders.

The Dee, 96 miles from source to sea, issues out of Braeriach, one of the Cairngorm summits, at the " Wells of Dee," and flowing eastward, enters Kincardineshire near Potarch. Through the three-spanned bridge of Potarch, between Aboyne and Banchory, it sweeps deep and strong over its gravelly bed. The road over Cairn o' Mount, the much-frequented old road from Tay

to Dee, formerly crossed here by a ford below the bridge.
For 12 miles the river continues its course through the
county, and then forms the northern boundary for the
remaining 14 miles. From Kincardine-O'Neil it receives
the Canny (9 miles) directly below Inchmarlo House
and close to Invercanny reservoirs, connected with the

Bridge of Feugh

Aberdeen water supply ; and also the small burns of
Cluny and Corrichie from the Hill of Fare. On the
south side the Feugh (20 miles), from the Forest of
Birse, flows for 8 miles to Whitestone. There it is
joined by the Aan (10 miles), which comes along the
county boundary from Mount Battock, and at Kirkton
of Strachan by the Dye from Glen Dye. The Bridge of
Feugh, 350 yards from the point where the Feugh and
Dee join, is one of the most noted and beautiful spots

on Deeside. The bridge itself, a plain structure, derives its picturesqueness from the rocky channel of the river both above and below it. Here the stream, embowered in a wealth of wood and greenery, courses swiftly over and around ledges of projecting rock, the foaming water, especially when the river is in flood, forming a magnificent spectacle. Though neither the longest nor the largest river in Scotland, the Dee lays claim to being one of the most rapid. Rising 4000 ft. above sea level and fed by numerous mountain streams, it has a flow of water remarkably pure, although it is subject to high and sudden floods. Its banks, throughout its entire course, are extremely well-wooded, while as a salmon stream it has few equals.

The North Esk, sometimes called the East Water, has a course of 40 miles to the sea from its source in Loch Lee in Forfarshire. Towards the foot of Glen Esk it touches Kincardineshire, and for the next 14 miles forms the south-west boundary of the county. For romantic beauty its course of 5 miles through the beautiful woods of the Burn could hardly be surpassed. Here the river has ploughed out for itself a deep gorge between rugged rocks, along which are pleasant winding paths, shaded by overhanging foliage. From the " Loup's Brig," as well as from Gannochy Bridge, a lofty arch, 30 feet high and 52 feet wide, on the Fettercairn to Edzell road, the foaming cataracts and dark pools confined between the rocks and cliffs form a scene of surpassing beauty and grandeur. The Esk drains an area of 224 square miles, 80 of which belong to Kincardineshire. The river has shifted its mouth several times in recent centuries, as is shown by an examination of the triangular patch of alluvial soil and sand, north of its present mouth. Up to the end of the eighteenth century it entered the sea 2 miles, and up to 1879 1 mile, further north than it enters now. The Luther (12 miles) is the

Gannochy Bridge

largest feeder of the North Esk from within the county. Encircling Strathfinella Hill, it flows first through the beautiful glen of Drumtochty, then sweeping southward and westward through the Howe, it joins the North Esk about 2 miles from the village of Marykirk.

Of the other streams in the east, the Bervie (16 miles) is the largest and most interesting. From the uplands of Glenbervie, it takes a south-easterly course, rounds Knock Hill, and winds through the fertile haughs of

B

Fordoun and Arbuthnott, reaching the sea at Bervie. Its banks are picturesquely wooded near the old mansion-house of Arbuthnott. The Bervie is a famous trouting-stream, and has a salmon-fishery at its mouth. The burn of Catterline in the east of Kinneff; the burn of Benholm in Benholm parish; and the burn of Finella in St Cyrus, enter the sea through gorges worn by the water in the rocks that crown their banks. The Den of Finella with its waterfall 70 ft. high rivals in grandeur the scenery of the North Esk at the Burn, though not on so extensive a scale. Tradition relates that, when Queen Finella was pursued after the murder of King Kenneth III., she fled here,

> "And leapt from the rocks to a wild, wild boiling pool,
>   Where her body was torn and tossed."

The lochs in the county are few and small. The Loch of Leys in Banchory was over 2 miles in circuit, but is now drained; Loirston Loch in Nigg is 27 acres in extent. There are two extremely pretty artificial lakes in the county. That within the policies of Fasque House is 20 acres in extent; and Glensaugh Loch, in the upper part of Drumtochty Glen, has certainly now little appearance of being artificial. Both have wooded islets, and are the haunts of wild duck and other water fowl.

6. Geology

From the point of view of origin, all rocks belong to one or other of two groups. There are the *igneous*

rocks, which have been at one time in a molten condi-
tion, and which have become consolidated by a process
of crystallisation ; while the *derivative* rocks, directly
or indirectly, result from the decay of pre-existing
rocks. Familiar examples of igneous rocks are the
lavas from modern volcanoes. Sometimes, however,
the molten matter fails to reach the surface, and is
consolidated, as granite for example, in or between
other rocks. It is then called *intrusive*. Derivative
rocks are often spoken of as *sedimentary*, because for
the most part they have been deposited as. sediments
in the flow of lake or sea. They may be recognised in
the field by their bedded or stratified character. Igneous
rocks, on the other hand, are unbedded. Many rock
masses have been so profoundly altered by heat, by
pressure, and by other causes, that their original characters are more or less obscured. Such rocks are
termed *metamorphic*. Examples of these are the widespread mica schists and gneisses.

   The deposits now forming in the sea floor tend to
be arranged in approximately horizontal layers. Very
often, however, as a result of coastal movements the
sedimentary rocks have been tilted (sometimes, as at
Stonehaven, the bedding planes are quite vertical) ;
or again they have yielded to pressure by folding or
fracturing. The folding may be simple, as in the rocks
which underlie the Howe of the Mearns ; or complicated,
as in the schists of the Grampians. A splendid illustration of a fracture or fault on a big scale is seen in
the " Highland Fault," which forms the geographical

boundary between the Highlands and the Midland Valley of Scotland. It enters Kincardineshire at the Woods of the Burn, and reaches the North Sea at Garron Point, near Stonehaven.

To many the chief interest of geology lies in the study of fossils, the remains of plants and animals preserved in the sedimentary rocks. Fossils enable us to ascertain the relative age of rocks and to classify them in groups and systems. The oldest rocks of the earth's crust, the Pre-Cambrian, contain few fossils. Overlying these are four great groups, which, taken in order of age, have been named as follows : (1) *Primary* or *Palæozoic* ; (2) *Secondary* or *Mesozoic* ; (3) *Tertiary* or *Cainozoic* ; (4) *Post-Tertiary*. The rocks of known age in Kincardineshire belong either to the Primary group or to the Post-Tertiary. The Post-Tertiary deposits include the boulder clays and fluvio-glacial gravels and sands, the raised beaches which fringe the coast, the alluvial terraces or haughs of the river valleys, and the peat mosses.

Considering first the solid rocks of the county, we find they are of markedly different character on opposite sides of the Highland Fault. To the north of that great fracture they belong mainly to the Dalradian series, to the south to the Old Red Sandstone. Between the Dalradian rocks and the Highland Fault, however, at the Woods of the Burn, at Glensaugh, at the Bervie Water, and at Elfhill, areas occur to which has been applied the term Highland Border rocks. On the coast between Cowie and Garron Point, but on the south side of the Highland Fault, rocks similar in their lithological characters have yielded fossils which indicate that they are in all probability of Cambrian age. Another interesting suite of rocks occupying the coast section from Ruthery Head to Stonehaven Harbour, and extending

Felsite Sill near Cove

inland for 7 miles, has recently been shown to contain characteristic Silurian fossils.

The Old Red Sandstone system of Scotland is subdivided into Lower, Middle, and Upper. Rocks belonging to the Lower series occupy most of the southern half of Kincardineshire. The Middle series is absent, and the Upper is found only in a narrow tract along the coast near St Cyrus.

The Dalradian rocks may be studied most conveniently

in the cliffs between Garron Point and the Bay of Nigg,
but numerous good sections are exposed in the streams
which traverse the hills between the valley of the Dee,
and the border of the Highlands. Intrusive rocks of
various types are found associated with the Dalradian
rocks. In the neighbourhood of Banchory, for example,
these have been "flooded" with a very old granite;
and later dykes are everywhere abundant. Further,
the dominating features in the scenery of the northern
half of the county are produced by intrusive rocks—
the "newer" granites on either side of the valley of
the Dee.

The Highland Border rocks consist of two groups:
an older series (probably Cambrian) made up of green
pillowy lavas, associated with red jaspers, green cherts,
and black shales; and a younger series of conglomerate
grits, limestone, and shales. Both groups show a
splendid development at the "Rocks of Solitude" in
Glenesk; and the fossiliferous shales of the older series
may be hammered in the cliffs at Craigeven Bay, Stonehaven. The most abundant fossils are early types of
Brachiopods or lamp shells. The limestone of the
younger series was at one time extensively worked.

During early Silurian times the region to the north
of the Highland Fault began to undergo compression
and elevation. The Dalradian rocks and the rocks of
the Highland Border series were thrown into great
folds; the coastal movements moreover heralded a
violent outburst of volcanic activity. We may picture
the Grampians of that period as a lofty mountain

range with numerous active volcanoes, snow-covered doubtless, and resembling perhaps the Andes of the present day.

The magnificent cliffs from Stonehaven southwards afford splendid opportunities for the study of the Lower Old Red Sandstone. Coarse conglomerates predominate, but occasionally give place to micaceous sandstones, while at intervals the succession of bedded rocks is broken by massive piles of lavas. Some of the bedded rocks, too, on close examination, prove to be volcanic tuffs, the consolidated " ashes " of the contemporaneous volcanoes. Tuffs occur also at Cowie, where their presence shows that volcanic activity had already begun in Silurian times. It continued until almost the close of the Lower Old Red Sandstone period. The hard resistant lavas form most of the high ground in the southern half of the county. The Garvock Hills, for example, are built up for the most part of a great succession of lava flows, and show beautifully from certain points of view the characteristic step-like arrangement which suggested the old name of " trap " rocks.

At the close of Lower Old Red Sandstone times coastal movements again made themselves felt in no uncertain fashion. The rocks of this period were compressed into simple " saddle-shaped " and " trough-shaped " folds—the Howe of the Mearns marks the position of one of the latter—and then, too, in all probability, was initiated differential movement along the line of the Highland Fault. The forces of denudation became active, and from the disintegration of the Lower Old

Red Sandstone and older rocks were built up the bedded rocks of the Upper Old Red Sandstone. The latter formation occurs in the coastal track between St Cyrus and the mouth of the North Esk, and is everywhere separated from the Lower Old Red Sandstone by lines of faulting. A vast epoch of time intervened between the deposition of the two formations. No fossils have been obtained so far from the Upper series in Kincardineshire, and the age of the rocks is inferred from their structural relations and from their lithological resemblances to fossiliferous rocks of like age in other parts of Scotland. One of the most characteristic rocks is a variety of nodular limestone known as "cornstone." This, like the limestones of the Highland Border, was at one time burned for lime.

Now follows, as regards our county, a great gap in the geological record. Of the story of the remainder of the Palæozoic epoch, and of the whole of the Mesozoic and Cainozoic times the rocks of Kincardineshire tell us but little, and that little very indirectly. In the Upper Old Red Sandstone period the highest forms of life were primitive fishes. Amphibians, reptiles, birds and mammals had, in succession, been evolved.

The Post-Tertiary deposits in Kincardineshire consist mostly of accumulations of sand and gravel, and of boulder clay or till with its characteristic striated boulders. They tell us of a time not so very long ago, geologically speaking, when the whole of Scotland, with the exception of a few of the highest mountain peaks, was buried deep in the ice sheet of the Great Ice Age.

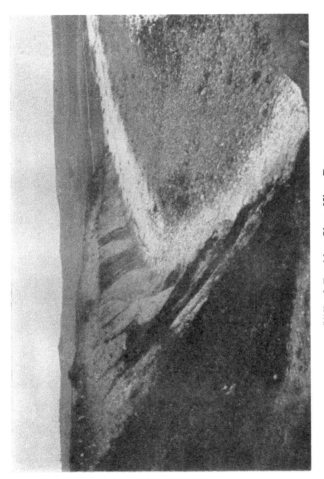

Cliff of Boulder Clay, Nigg Bay

The striated stones are the tools with which the ice sheet accomplished its work. How effectively that work was done is evidenced by the rounded, flowing contours of our hills, by the presence of boulder clay and erratic blocks, by the glacial grooving on a big scale

Striated Stone from Nigg Bay

wherever belts of soft rock lay in the path of the ice, and by the preservation of the ancient bottom moraine, the great thickness of till which conceals the solid rocks over much of the county. That the minor surface features are largely glacial in origin cannot for a moment be doubted. One instance must suffice. No one travelling along the Howe of the Mearns can fail to note the contrast offered by the bordering hills. On the one side, the even boulder-clay-covered slopes of the Garvock Hills rise gently from the plain ; on the other, every valley opening from the Grampians is fronted by one or more steep-faced terraces. The terraces consist of sand and gravel deposited in lakes formed at a time when, while the local hills were free from ice, a great lobe of the Highland ice sheet still occupied the Howe. Similar phenomena are seen in the wide valley of the Dee.

7. Natural History

In recent times—recent, that is, geologically—no sea separated Britain from the Continent. The present bed of the North Sea was a low plain intersected by streams. At that period, then, the plants and the animals of our country were identical with those of Western Europe. But the Ice Age came and crushed out life in this region. In time, as the ice melted, the flora and fauna gradually returned, for the land-bridge still existed. Had it continued to exist, our plants and animals would have been the same as in Northern France and the Netherlands. But the sea drowned the land and cut off Britain from the Continent before all the species found a home here. Consequently, on the east of the North Sea all our mammals and reptiles, for example, are found along with many which are not indigenous to Britain. In Scotland, however, we are proud to possess in the red grouse a bird not belonging to the fauna of the Continent.

The flora of Great Britain has been divided, as regards

climatic types, into four classes—(1) Alpine ; (2) Sub-Alpine ; (3) Lowland ; (4) Maritime. Kincardineshire, with its diversified soil and situation and with an elevation reaching over 2000 ft., has representatives of all the four classes. The county as a whole is remarkably rich in the number and the variety of its wild plants, while several spots within it have acquired more than a local reputation as a hunting ground for the botanist. Thus, we have on the coast the well-known St Cyrus braes, where, owing to favourable conditions, a large number of plants occur that are not found in other parts of the county. Here the volcanic rocks decompose into a light brown soil, extremely suitable for the growth of wild flowers, unless when exposed to continuous drought, which in our climate does not often occur. The exposure of the rocks, forming cliffs almost 200 ft. high, facing south and east, adds to the warmth afforded by the soil. Here, during the summer, may be seen in abundance the pretty little maiden pink, the prolific rest harrow, bladder campion, viper's bugloss, bloody crane's bill, hemp agrimony, common cudweed (the *herba impia* of old writers), butterbur, marjoram, goat's beard, red poppy, field pepperwort, soft knotted clover, rough podded yellow vetch, field garlic, wild sweet pea, Nottingham catch-fly, and others.

On the loose sands along the banks of the North Esk and in the salt marshes at its mouth grow the lesser meadow rue, the sea rocket, the thrift or sea pink, the prickly saltwort and other similar plants. Close to the river, on ground liable to be flooded at high tides, may be found sea pearlwort, sandwort spurreys, sea milk-wort, jointed glasswort, sea arrow grass, and several varieties of sedges. Grass wrack, one of the few flower-ing plants of salt water, grows in the mud at the old mouth of the river.

The braes and seashore of Muchalls, though inferior to St Cyrus in number and variety of specimens, are of great interest to botanists. Besides some of the commoner plants already mentioned, Muchalls supplies lamb's lettuce, sea wormwood, white campion, the lovely oyster plant, and several varieties of worts, willow herbs, vetches and sedges.

Inland, wild flowers abound, especially along the rivers, in the sheltered glens, and on the wooded hills. To mention only a few, we have lung wort, wintergreen, cordalys, wood bitter vetch, celery-leaved crowfoot, buckbean or bogbean, water plantain, comfrey ; with such commoner forms as ragged robin, greater and lesser celandine, lady's bedstraw, knapweed, golden rod, eye-bright, field gentian, forget-me-not, and ground ivy. The Alpine flora includes Alpine lady's mantle, willow herb, mountain and water avens ; and in the bogs and marshy slopes, sundew (*Drosera rotundifolia* and *D. latifolia*), very plentiful in Netherley moss, butter-wert, bog orchis, bog violet, and others.

Numerous varieties of the mosses of north-east Scotland occur in Kincardineshire, in the wet and boggy parts by river banks, as at the Burn or along the Dee valley, and at the seaside. Of ferns, beside the common polypody, which is very abundant, we find the beech fern, the graceful oak fern, five varieties of the Aspidia (including the rough Alpine shield and the close-leaved prickly shield), the bladder fern. Five or six spleen-worts (including the wall rue spleenwort and the sea spleenwort) grow along the cliffs in the south of the county and as far north as Muchalls and Portlethen, where also the black spleenwort has been gathered.

The fauna of Kincardineshire includes the ordinary animals of the country. The fox is not so numerous as he was in the early years of the nineteenth century, when fox hunting, now entirely given up in the shire, was indulged in by some of the county gentlemen. The brown hare has, however, increased very much in numbers, not always to the advantage of the farmer's crops. The blue or mountain hare is plentiful on the Grampian slopes. Wild rabbits, known only as children's .pets in the county before 1808, abound everywhere. The otter is occasionally seen by the side of the larger streams, but the badger and wild cat are now extinct. Squirrels, unknown in the county a century ago, are now fairly numerous. Roe deer are found in the lower Grampian slopes, and red deer sometimes in the Glen of Dye and elsewhere. Grouse and partridge are numerous, while the heron builds in the high trees by the North Esk and the Bervie, and may be seen feeding in the river pools. The capercailzie has come by Dye and Feugh to Dee.

To the sea birds the cliffs afford a secure retreat and a fitting nursery. The guillemot, credited errone-ously with being a stupid bird, asserts his superiority in number over the kittiwakes, tommie-nories, or Greenland parrots, gulls and coots, which inhabit the precipitous ledges of their summer home.

Gulls' Crag, Stonehaven

With the increase of woods and other shelter the smaller birds have increased in number. The yellow-hammer, hedge sparrow, chaffinch, stonechat, and other similar birds are everywhere in evidence by the roadsides and fields, while the blackbird, the starling, and the mavis are not averse to sampling the products of the fruit garden in summer or early autumn.

The goldfinch and the siskin are now extremely rare, while the magpie is decreasing in numbers. The ptarmigan is extinct. The golden eagle is practically extinct, though one or two have been sighted in the hills above Drumtochty. Rare visitants are the quail, snow-bunting, great spotted woodpecker, Bohemian waxwing, little auk, Manx shearwater, hen harrier, peregrine falcon, and common buzzard ; but these can only be regarded as accidental visitors, driven thither by stress of weather or other circumstances.

8. The Coast

The seaboard, 35 miles in length, which Kincardine-shire possesses, is perhaps as interesting as any other part of the Scottish coast, on account not merely of its picturesque rock scenery but also of its historical asso-ciations. All the way paths run close to the sea, from many points in which splendid views can be got of maritime and inland scenery, though undoubtedly we obtain the best idea of its beauty when sailing along the coast. Like most of the eastern seaboard of Scot-land, the Kincardineshire portion is much exposed to the strong gales sweeping in from the North Sea ; and this, combined with the rocky nature of the greater part of the shore, accounts for many of the shipping disasters that occur.

Girdleness Lighthouse

c

Starting our peregrination from Aberdeen we note first the Bay of Nigg, flanked on the north by the lighthouse of Girdleness, and on the south by Greg Ness, the circular outline of the bay being fringed by a beautiful pebbly shore. Here, formerly, was the

Low Tide at Nigg Bay
(*Showing stones from cliff of boulder clay*)

mouth of the Dee, which flowed in the hollow from Craiginches. On the bay stands a fish-hatchery with laboratory. A little inland is St Fittick's ruined church. Prominent on the south of the bay is a cliff of boulder clay, the rapid erosion of which has littered the beach with thousands of stones.

Passing on, we find the coast bold, rocky, and pictur-esque ; and we reach in succession the small fishing

The " Old Man " of Muchalls

villages of Cove and Portlethen. Between them, but back from the cliffs, is Findon, world-famous as the original home of the "Finnan haddock." "The haddocks cured there," says Thom (*History of Aberdeen*), "are superior in flavour and taste to any other, which is attributed to the nature of the turf used in smoking them." The industry is now entirely given up in Findon. Skateraw, a little further south, is, like the other creeks, reached by a narrow, circuitous path down the sea slopes, up which in former days the hardy fishermen carried in their creels the shining "harvest of the sea" to be transported by road or rail to the larger centres of population. Part of the fish supplies landed here were split and sun-dried on the stony beach, and went by the name of "speldings." Like the "Finnan haddie," these, when properly cooked, were held in high esteem. The small burn of Elsick, spanned by a substantial railway viaduct, here enters the sea.

The next part of the coast, adjacent to the neat little village of Muchalls, has received much attention from the painter of maritime subjects, and deservedly so, because of the artistic beauty of the rugged, weather-beaten cliffs. Here by the ceaseless action of the elements the softer portions of the cliffs have been scooped out into long, deep gullies through which in stormy mood the sea rolls with resounding and majestic grandeur. The "Fisher's Shore," the "Grim Brigs" with its wonderful arches of Nature's own devising, the "Old Man," and the "Scart's Crag," around and above which for ever breaks the crested wave, are notable points whose names, like that of " Gin Shore," a little further south, are reminiscent of the past, and full of interest and suggestion.

Between Muchalls and Stonehaven we pass Garron Point, on whose green summit stand the picturesque ruins of the old chapel of Cowie. The little fishing

Dunnottar from the North

village of Cowie nestles below the cliffs, while above, skirting the shore, is the Stonehaven Golf Course, from which splendid views can be had of sea and shore.

Stonehaven Bay extends in a circular sweep from Garron Point to Downie Point. Alongside of its pebbly beach runs a promenade, flanked on the north by extensive recreation grounds. Its waters give ample scope for bathing and boating, while the dull grey and brown outlines of the Old Town dwellings at the southern end impart an old-world appearance to the scene.

Rounding the Black Hill, from the top of which
unrolls one of the finest views of town, coast and inland,
we reach the historic castle of Dunnottar, where, as
Carlyle's eulogy of the famous Marshal Keith reminds
us, "The hoarse sea winds and caverns sing vague

Fowlsheugh

requiems to his honourable line and him." Here the
panorama formed by cliffs and bay is magnificent—the
former almost 170 ft. high with cathedral-like arches,
the latter with gloomy creeks and caverns. The very
names, as "Brun Cheek," "Maiden Kaim," "Long
Gallery," "Wine Cove," testify to Nature's handiwork
and skill. South of Trelung Ness we reach the highest

of the rocks, the cliffs of Fowlsheugh, the noted nursery
for sea birds, extending over a mile.   The birds make
their nests in the crevices of the conglomerate rock,
out of which by constant weathering pebbles have been

The Great Cave, Fowlsheugh
(*Looking out to sea*)

forced, affording a natural nesting-place.   The spectacle
of the myriads of birds in early summer, on

> " the dreadful summit of the cliff
> That beetles o'er his base into the sea,"

is most interesting and instructive, and will well repay
a visit from others than bird-lovers.

Between Fowlsheugh Point and Bervie Bay the cliffs
are still bold and precipitous, with generally no beach
between their base and the deep water.   Todhead light-

house stands on a prominent headland at the southern extremity of Braidon Bay, a little beyond the old fashioned fishing village of Catterline. Craig David, a few miles further south, overlooking Bervie Bay, marks the terminus of the high cliffs which form the natural wall of protection to most of the Kincardineshire coast.

From this point onwards the configuration of the coast-line is entirely different. The beach is now low, pebble-strewn, and gravelly, with low, shelving rocks jutting out to the sea. Gourdon—dominated by Gourdon Hill, a noted landmark for seamen—and Johnshaven have both small harbours, their appearance from above being quaint and picturesque. A little further south is the hamlet of Milton of Mathers. Rounding a bend in the coast we pass the Kaim of Mathers, and reach St Cyrus braes, varying in height from 50 to 300 ft. On the summit stands conspicuous the parish church with its lofty spire. Passing over a flat beach of fine sand bound together by sea grasses and other marine plants, we end our perambulation of the coast at the mouth of the North Esk.

The coast-line of the county bears witness to the gigantic power of marine erosion. Cliffs and bays, caves and half-tide stacks, show that the action of the sea in sculpturing coastal scenery is everywhere guided by rock-composition and structure. In this connection we may quote the words of Sir Charles Lyell with regard to a case of historic interest. " On the coast of Kincardineshire an illustration was afforded, at

Sea Cave near Cove

the close of last century [the eighteenth], of the effect of promontories in protecting a line of low-shore. The village of Mathers, two miles south of Johnshaven, was built on an ancient shingle beach, protected by a projecting ledge of limestone rock. This was quarried for lime to such an extent that the sea broke through, and in 1795 carried away the whole village in one night, and penetrated 150 yards inland, where it has maintained its ground ever since, the new village having been built further inland on the new shore."

In late glacial and post-glacial times there took place alterations in the relative level of land and sea. The raised beaches of the coast and the alluvial tracts of the river valleys were formed when the land stood relatively lower than at present. Two sets of beaches are clearly marked in Kincardineshire. Both are well seen at Stonehaven, the newer part of which is built on the flat of the 100-ft. beach, the older on the 25-ft. beach. The lower beach shows its best development southwards from Bervie, the old sea-cliff forming a strong feature all the way to the mouth of the North Esk, and the flat rocky foreshore of the present sea margin offering a striking contrast to the frowning cliffs which bound the shore from Bervie to Stonehaven harbour. That the land at one period stood higher (or the sea lower) than at present is shown by the occurrence of a buried forest beneath the 25-ft. beach.

9. Climate and Rainfall

The climate of a district, which may generally be defined as its average weather, depends upon the amount of heat and moisture, and these in turn depend upon latitude, altitude, slope, and, in some degree also, on the state of cultivation. Britain is in the same latitude as ice-bound Labrador, yet it possesses a temperate climate, due to the fact that the prevailing winds being from the west or south-west bring with them a certain amount of moisture and heat acquired in their passage over the Atlantic, which is three degrees warmer than the air. This explains why the west coast of Britain is warmer than the east. For the same reason the east is drier than the west, for the winds in passing over the mountains part with their moisture before reaching the east coast.

The physical configuration of Kincardineshire has much to do with both its temperature and its rainfall. Situated on the extreme south-eastern slope of the central highlands, it has behind it the immense extent of the Aberdeenshire, Perthshire, and Argyleshire mountains, shutting it off from the Atlantic seaboard. Thus the winds of winter from that quarter get gradually cooled and, reaching the east, speak not of the warm Atlantic, but of the snow-clad Grampians. One degree of diminution of temperature for every 300 ft. of elevation brings the west-coast temperature of 39° for January down to 29°, or 3° below freezing point, at the ridge of the Grampians, an elevation of almost

3000 ft. above sea-level. There is, even at sea-level, a difference of from 2° to 3° between the mean annual temperature of the west and the east coast.

The mean annual temperature of the county is 46°; of summer 58°; and of winter 37°. For comparison we give the temperature at Cowie Mains, Stonehaven, for four years:

Year	Average Maximum	Highest Maximum	Average Minimum	Lowest Minimum
1909	50.08	64.58	37.62	31.79
1910	50.95	62.2	40.45	29.3
1911	56.5	60.25	41	31.37
1912	56.1	62.16	40.8	30.58
Average	53.4	62.29	39.97	30.76

During the winter the greatest amount of snow is from north east and east, the most intense colds are from north and north-west, and the greatest amount of heat from south and south-west. The fact that Kincardineshire lies open and shelterless to the North Sea accounts for the biting winds of early spring, often accompanied by heavy rains. From the same source very often in April and May chill haars and hazes set in towards evening. Hoar frosts are prevalent in the neighbourhood of the mosses and low-lying marshy spots in the county, although by improved drainage and cultivation the area subject to this has been materially reduced.

Kincardineshire, being much diversified into hill and dale, with a great variety of altitude and exposure, has a difference of climate in its various divisions. Deeside, for example, although in the extreme north of the county, is the warmest district. This arises from several causes. First, it has a genial southern exposure. Secondly, it is sheltered by a number of small hills rising gradually from the Dee. Its dry gravelly soil readily absorbs any excess of moisture, while its pine woods and thriving plantations moderate the climate as well as adorn the landscape.

The climate of the Howe is both warm and equable. On the north the Howe is protected by the Grampians ; on the south and the east it is sheltered by the Garvock Heights from the full sweep of the North Sea winds in winter and early spring ; at the southern extremity it is open to the genial westerly breezes of summer.

In the heat of summer the coast is delightfully cool and refreshing ; but the glens and hollows of the Grampians are often very close and warm, though extremely cold in winter.

On the whole, the climate of Kincardineshire is bracing and healthy, with remarkably pure and exhilarating air.

The rainfall of the county, compared with the excessive fall on the west coast of Scotland, is relatively small. The mean depth in inches is 32.25, as against 40, 45, or even 70 inches on the Atlantic coast, and 44 for the whole country. At Stonehaven the average annual rainfall for the past twenty-one years has only been 27.13 inches ; which partly explains its popularity

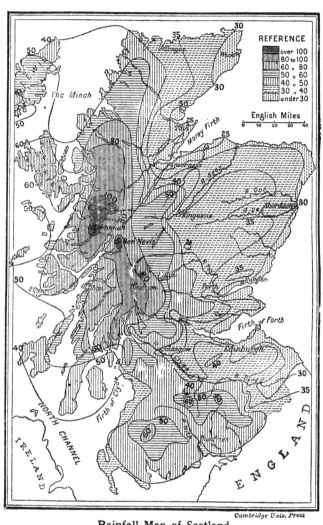

Cambridge Univ. Press

Rainfall Map of Scotland
(*By Andrew Watt, M.A.*)

as a holiday resort.   During the same period the average
for any month of the year has not been more than
3 inches of rainfall, the average for the first four months
of the year being 1.86 inches.   An increase in elevation
usually brings with it an increased rainfall.   Thus the

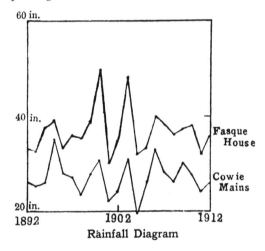

Rainfall Diagram

Burn, 14 miles inland, with an elevation of about
300 ft., has a fall of over 35 inches ; Banchory (almost
200 ft.) 30 inches ; Fettercairn (230 ft.) 32 inches.
The diagram, above, of the rainfall for twenty-one
years, from 1892 to 1912 inclusive, at Cowie Mains,
Stonehaven, and at Fasque House (330 ft. above
sea-level), illustrates this, besides showing the annual
variations in rainfall at each of these places ; and may
be regarded as fairly typical of the maritime and the
inland districts of the county.

10. People—Race, Language, Population

The Alexandrine geographer Ptolemy gives the dwellers between Dee and Tay the name of Venicones. These were part of the race of Picts, who occupied Eastern Scotland from the Pentland Firth to the Forth. Through the district now called Kincardineshire ran the dividing line between the Northern and the Southern Picts—the Grampians. Gaels also from the west found their way into this region.

Traces of the Pictish and the Gaelic occupation are discernible in place names. "There is no district," says Dr Don (*Archæological Notes on Early Scotland*), " in which Scottish land names may be better studied than in the ancient and still linked provinces of Angus and Mearns . . . they hold almost every type of Celtic and Saxon place name found in the country." *Pit* or *pet* and *fother* are Pictish, as *Pitnamoon, Pitforthie, Pitskelly, Pitgarvie, Pitbeadly, Fordoun, Fettercairn*. Of Gaelic origin are names of rivers, as *Esk, Bervie, Aan, Cowie, Luther*; of mountains, as *Clochnaben, Kerloch, Cairnmonearn, Knock, Carmont, Bruxie*; as well as *Kincardine, Mearns*, and the names of many of the parishes.

Towards the end of the fifth century the English invasion began. Over the North Sea strangers came sailing from Frisia and the adjoining districts to settle along the coast and originate the fishing villages. From these settlers, who in time pushed inland and inter-married with Picts and Gaels, the bulk of the people have sprung. This blending has produced the robust type of character that distinguishes the inhabitants to-day. Place names indicating English settlements are those ending in *ton, ham* or *hame, kirk*.

It is doubtful if any Norsemen made their homes here. But we find *ness*, from a Norse word for headland, in *Girdleness, Greg Ness*.

The Celtic tongue formerly spoken in Kincardineshire retreated long ago before a variety of Northern English. Gaelic speaking is now extinct, though at the census of 1911, 78 persons were recorded as able to speak Gaelic and English.

The vernacular of the county belongs to the Northern Division of the Scots dialects (extending along the east from the Tay to Caithness), but it has a few Midland characteristics. In pronunciation, for example, while in the regions towards the Dee words like *moon, school, good* are sounded in the northern way as *meen, skweel, gweed*, in the south they have the *ui* vowel sound, something akin to the sound in French *mur, peu*. The change of *wh* to *f* (characteristic of the old Pictish region) is in Kincardineshire still heard, but mostly in *fa, fat, fan =* *who, what, when*, and such like. The vowel sound in the pronunciation of *one, bone, stone* is as in the Aberdeenshire *een, been, steen*. Stonehaven is locally known as Steenhive. Unheard north of the Dee is the pronunciation of *knock, knee*, as *tnock, tnee*. This links the dialect with Forfarshire, and reminds one of J. M. Barrie's *Tnowhead* for *Knowhead*. It may be also noted that the forms *this* and *that* are plural as well as singular.

D

This steens, that beens are *these stones, those bones.*
Dialect differences, however, are to a certain extent
disappearing under the influence of schools, newspapers,
and easy communications.

As regards population Kincardineshire with 41,007
inhabitants stands twenty-fourth in the list of Scottish
counties. Since 1801, when the first census was taken,
there has been an increase over the whole county of
14,659, or 55 per cent. From that date each decennial
census has shown an increase with the exception of
those of 1861 and 1881, when the decreases were very
small. The relatively great increase in the 1901 returns
(14.8 per cent.) is explained by the fact that 11,428 were
included in the Kincardineshire returns as the population of Torry, which really forms part of Aberdeen city.
With this excluded, the population of the rest of the
county is found to have decreased by 1957, or 6.2
per cent.

11. Agriculture

The high position of agriculture in Kincardineshire
cannot be thoroughly understood without a reference
to the enthusiasm for improvement displayed by many
of the landed proprietors in the latter half of the
eighteenth century. Of these the most remarkable
was the famous agriculturist, Barclay of Urie, whose
work, as pointed out in Robertson's *Agricultural Survey*,
reads like a romance. In the half century that followed
the Union of 1707, he had acquired, from residence in
England, very advanced ideas in agricultural theory

and practice ; and not only did he translate these into practical reforms, but he also infected others with his own keen desire for agricultural advancement. When he entered into possession of Urie, it was " a complete waste, consisting of bogs, baulks and rigs, everywhere intersected with cairns of stones and moorland." And this description might, with even greater truth, have been repeated of most of the large estates in the county at that period. Such was his energy that in the short space of thirty years he materially improved 2000 acres, reclaimed from moorland 800 acres, and planted from 1200 to 1500 acres with trees, an evidence of the latter work being still seen in the magnificent woods of Urie.

The methods he employed were both intelligent and skilful. Throwing up, by trench ploughing, the incredible quantity of stones which lay in the soil, he utilised thousands and thousands of tons of these in making drains and dykes. The land thus improved was treated to a liberal supply of lime, to render it more productive and kindly, and better suited for growing turnips and artificial grasses, which he introduced into crop rotation, a system up to that time sparingly practised.

The neighbouring proprietors followed his example, and the closing years of the eighteenth century thus became remarkable for agricultural progress in Kincardineshire. Since then vast changes have taken place in agricultural theory and practice. Communication between farm and market has been made much easier by the introduction of railways and the improvement of roads—a circumstance which has indirectly led to the abolition of fairs and markets. Labour-saving machinery has been introduced ; scientific methods are now adopted in the culture, manuring, and draining of fields ; in the rearing, feeding, and general treatment of his stock, the farmer has at his command to-day the very best results of scientific experiment and research.

In Kincardineshire mixed farming is general. On the hill grazings of the Grampian slopes, more attention is naturally paid to sheep-rearing than to tillage ; but even on these farms all the available land is reserved for cereals or grasses.

The area of the county is 248,195 acres, of which 127,923 acres are waste or heather, not under the plough, which leaves about 48 per cent. of cultivated land, as against 24.2 for the whole country.

Twenty-first in area and twenty-fourth in population among Scottish counties, Kincardineshire stands in acreage under cultivation as follows : for barley 7th, for turnips 9th, for potatoes 11th, for wheat 12th, for oats 16th. The high position in regard to turnips is because the county is a feeding as well as a breeding area for cattle and sheep. Practically one-sixteenth of the whole barley acreage for Scotland is in the Mearns, the soil of which is remarkably well adapted for the growth of barley. Of the 708 acres of wheat grown in the county in 1913 more than half was grown in the strong lands of the St Cyrus district, the remainder in the Howe, chiefly around Laurencekirk ; while on Deeside with its light gravelly soil it was entirely absent. The cultivation of oats, potatoes, and turnips is well distributed over the county.   Only a few acres are given to rye and beans.

The county does not, like Aberdeenshire and Forfarshire, possess any distinctive breed of cattle ; but among the early improvers of cattle breeds were several notable Mearns men ; and to-day the Burn and the Portlethen herds are well known to agriculturists.

The following is a comparative table of the number of the live stock in the county at the beginning of last century and in the years 1913 and 1917 :

	1807.	1913.	1917.
Cattle . .	24,825	27,731	24,717
Sheep . .	24,597	47,694	48,293
Pigs . . .	478	2,139	1,816
Horses . .	2,579	4,700	5,346

Frequent mention is made of the abundance of timber in Kincardineshire in early days ; and the existing plantations show the suitability of certain districts for the growth of forest trees.   On the Durris estate some of our exotic trees were first introduced, and have given the most remarkable results.   Two species have here shown their superiority—the Douglas fir and the Menzies spruce.   The former, owing to its free growth, freedom from disease, and wonderful adaptability to a wide range of soils, has proved itself capable of producing more volume per acre than any other species of exotic tree.   One Durris specimen of the Douglas fir, measured in 1904, was 110 ft. high.   The whole of the Deeside

In the Birkwood, Banchory

district is, however, well suited for the growth of timber, the other principal forest regions being along the southern spurs of the Grampians.   In many cases the lower hills are wooded to the summits.   The usual trees grown are Scots fir, larch, spruce, and the commoner hardwood trees—-ash, plane, elm, beech, birch, and oak.

12. Manufactures and Other Industries

According to the last census returns, five out of every twelve of the adult population of Kincardineshire were directly engaged in agriculture ; but if we consider those indirectly engaged in it and in its allied occupations, the proportion would be almost doubled.   Other industries, then, take a secondary place.   In the absence of large towns to attract the rural population, there is not much concentration of labour nor any great development of the factory system, as in Forfarshire.

The first linen-yarn mill in Scotland was set up at Bervie in 1790 ; and flax spinning, formerly an important home industry, is still carried on at Bervie, as well as at Gourdon and Johnshaven.   Handloom weaving was a widespread occupation in most of the towns and villages till steam power was introduced about 1850, when many weavers found employment on the infant railways.   Handloom linens are still made in Laurencekirk, but elsewhere the industry is extinct. Stonehaven has a mill for woollen fabrics and hosiery, and a flourishing factory for fishing-nets.

There are distilleries at Glenury, Fettercairn, and

Auchinblae ; and a brewery at Laurencekirk. The development of the bicycle and motor-car industry has, in recent years, given employment to an increasing number of skilled workmen in the county. Laurencekirk and Stonehaven are centres for carriage building. The well-wooded valley of the Dee has several sawmills, supplying pit-props for mining districts and timber for box- or case-making in Aberdeen and elsewhere.

A manufacture, long extinct, was the making of a special kind of snuff-box in Laurencekirk. The peculiarity of the box was a concealed hinge and pin, invented by Charles Stiven about 1780.

Kincardineshire has neither coal nor iron ore. In the end of the eighteenth century large quantities of an irregular mineral substance called " native iron " were, however, found in Fettercairn. Detached pieces of various sizes were turned up by the plough, which were converted into use by heating and hammering in the local smithies. The origin of this metallic substance, which was soon exhausted, was never properly accounted for, although many theories, fantastical and otherwise, were propounded.

Granite is quarried at Cove and Hill of Fare. Formerly this industry seems to have been of more importance in certain parts of the country than it is now. At the beginning of last century, for example, about 600 hands were employed in the Nigg quarries. From these, granite blocks, squared and dressed, were shipped at Aberdeen to pave the London streets. Sandstone is freely distributed over the county, and much of it is utilised for road metal. The quarries of Lauriston, St Cyrus, and Threewells, Bervie, supply excellent building-stone, which is easily wrought.

Another industry, now entirely given up, was limestone burning. The lime from the kilns of Mathers, St Cyrus, was in great demand among farmers. Similar kilns existed in Fordoun, Fettercairn, and Banchory. Parts of old kilns still remain at Clatterin Brigs and Mains of Drumtochty.

13. Fisheries

Britain being an island surrounded by shallow seas in which fish are plentiful, it is only natural that the fishing industry should be one of the most important sources of wealth as well as of food. Fishing is carried on vigorously on both the west and the east coast of Scotland, but the east coast fishing is of far greater magnitude and importance than the west coast. The North Sea is not only an excellent fishing ground, but it also has splendid ports where the catches can be disposed of to advantage. The following returns of the value of the fish caught on both coasts for 1912 bring out their relative importance :

East Coast—Total value of all fish landed	£2,323,580
Orkney and Shetland— ,, ,, ,,	775,209
West Coast— ,, ,, ,,	352,040
Grand Total	£3,450,829

The value of shell-fish caught on the west coast, however, exceeds considerably that on the east coast.

The whole country is divided into districts by the Fishery Board for Scotland each district being in charge of an officer, whose duty it is to get and to give information on all matters connected with the industry. The Kincardineshire small ports or "creeks," as they

Herring-boats, Gourdon

are called, are connected with the three districts of Aberdeen, Stonehaven, and Montrose. Downies, Port-lethen, and Cove are naturally linked on to Aberdeen ; Milton, Johnshaven, and Gourdon to Montrose. Stone-haven includes Cowie and Skateraw to the north, and Catterline and Shieldhill to the south.

The chief kind of fish landed on the Kincardineshire coast in 1912, arranged in order of market value, were herrings, codlings, haddocks, whitings, crabs, lobsters, which with less important varieties reached a total value of £21,329, almost one-eighth of this being the value of the shell-fish caught. The weight of all the fish landed (excluding shell-fish) was a little over 2000 tons. Between five and six hundred fishermen are engaged in the industry, while 235 boats or vessels of various sizes belong to Kincardineshire.

Since 1902, when motor power was first introduced into the fishing industry, the progress and increase of motor boats, slow at first, has been very marked. In this innovation, the pioneers in the county were the fishermen of Gourdon and Johnshaven.

More than a century ago salmon fishing gave employment to 135 hands, and the rental of the fishings amounted to £2700 a year. At present the assessable rental of the three districts—Bervie, North Esk, and Dee—is £27,825, about one-sixth of the rentals of the forty Scottish districts having boards to regulate and protect salmon fishing. The number of salmon caught annually either in the sea by a " fixed engine "—the stake and bag nets—or by rod in the waters of the rivers of the county, cannot be determined ; but the weight of salmon carried by the railways in 1912 was 1990 tons, almost half of this quantity being caught in the area from Berwick to Cairnbulg Point in the north-east of Aberdeenshire. Of this a considerable proportion must have been contributed by the Kincardineshire salmon fishings.

14. History of the County

The history of the county, though interesting, has not been much concerned with the great events of national history. And yet the existence in early days of a royal residence at Kincardine indicates a certain importance. Kincardine was probably chosen as a residence by the Pictish kings, because it commanded the pass of the Mounth and the road to the eastward. Its castle may have dated from the reign of William the Lyon. In mediaeval times it was one of the chain of strongholds guarding the route from Forfarshire over the Mounth to the north—Brechin, Kincardine, Loch Kinnord, Kildrummy, Strathbogie, Rothes, Elgin, Duffus, Blervie, Inverness, Dunskaith. As a royal residence, it grew less important when the midland centres increased in power and influence, and it ceased to be the capital of the shire in 1600, when Stonehaven became the chief seat of local administration.

That the Roman legions under Severus (A.D. 208) passed through the county is undoubted, though the events connected with this invasion are obscure and disputed. Goaded into revenge by the insurrections of the wild Caledonians, he set out himself with a strong force, and at once began the formation or the continuation of the road through the north-eastern lowlands. The route of the Roman armies through Strathmore and the Mearns is clearly mapped out in the sites of the camps which run in a line from Tay to Dee. These were at intervals of about 12 miles, or a day's march ;

and it is reasonable to assume that of the 50,000 soldiers lost by the Emperor in his Caledonian campaign, a certain proportion must have fallen in the conflict with the sturdy " Men of the Mearns." The Roman camps in the county are said to have been at Fordoun, and Raedykes, near Stonehaven, while Normandykes, in Peterculter, is just beyond the county border. This view, strongly held by some authorities, is strongly condemned by others. The battle of the camps will have to be decided, if that is now possible, by excavations on the sites. It is noteworthy that, in the Raedykes–Normandykes area, Roman relics have been unearthed—coins, swords, pots.

In the wild days when Scotland was in the making, when Picts and Scots, Angles of Lothian and Britons of Strathclyde, struggled for mastery, the Mearns on the route from Fife and Perth to Aberdeen and Moray must have been the scene of many a bloody conflict. After the union of Picts and Scots, Kenneth MacAlpin's immediate successors found the Mearns a constant source of trouble : it was there that three kings died a violent death. In 954, Malcolm I. was defeated and slain at Fetteresso, though some say he was killed in Morayland. Forty years later Kenneth III. incurred the enmity of Finella, wife of the Mormaer of the Mearns, whose son had died in battle against the king. By her contrivance Kenneth was killed, but how is not certain. Hector Boece's account is grimly picturesque. Kenneth had visited Finella's castle at Fettercairn and was conducted into a tower, " quhilk," to use the words of

Bellenden's Scots version of Boece, " was theiket with
copper, and hewn with mani subtle mouldry of sundry
flowers and imageries, the work so curious that it ex-
ceeded all the stuff thereof." There stood a statue of

Sculptured Stone, Fordoun
(Supposed to commemorate Kenneth III.'s murder)

the king, in his hand a gem-studded apple of gold.
The apple (so Kenneth was told) was a gift for himself.
Would he deign to accept it from the hand of the image ?
He touched the apple, and at once a shower of arrows
pierced his body. In 1094, when rivals claimed Malcolm

Canmore's throne, the Mormaer of the Mearns, Malpeder MacLoen, backed Donald Bane against Duncan II. In a battle at Mondynes in Fordoun parish, Duncan died. A great stone on a knoll in a field, called Duncan's Shade, is believed to commemorate the spot.

In common with the other parts of the east coast, Kincardineshire suffered from the inroads of the Danes during the tenth and the early part of the eleventh century. At the battle of Barry, their leader, it is said, was killed by the founder of the Keith family, and was buried at Commieston in St Cyrus.

During the period of the Wars of Independence Edward I. passed through the Mearns on his triumphal march northwards (1296). From Montrose he directed his course to " Kincardine in Mearns Manor," then to Glenbervie Castle, where he stayed a night, next over the Cairn O' Mount to " Durris manor among the mountains." According to Blind Harry, Wallace over-ran the Mearns in the following year, and penned 4000 Englishmen within Dunnottar.

> " Wallace in fyr gert set all haistely,
>   Brynt wp the kyrk, and all that was tharin,
>   Atour the roch the laiff ran with gret dyn.
>   Sum hang on craggis rycht dulfully to de,
>   Sum lap, sum fell, sum floteryt in the se.
>   Na Sotheroun on lyff was lewyt in that hauld,
>   And thaim within thai brynt in powdir cauld."

In 1562 the battle of Corrichie was fought on the south-east slope of the Hill of Fare. Queen Mary was making a progress through the northern shires when the Earl of Huntly turned rebellious. The royal forces, under the Earl of Moray, defeated the rebels at Corrichie. From a spot still named the Queen's Chair, tradition says Mary viewed the fight.

In 1639 the Marquis of Montrose and his men passed through the county on their way to Aberdeen to compel

Old Bridge of Dee

the people of Aberdeen to sign the Covenant. The Earl Marischal and other " Men of the Mearns " joined him. During the operations round Aberdeen occurred the " Raid of Stanehyve." Viscount Aboyne crossed the Dee with 2500 men, plundered Muchalls and had reached Megray Hill, close to Stonehaven, when their opponents met them, well supplied with cannon from Dunnottar. Highlanders feared " the musket's mother," as they designated the cannon ; and those in Aboyne's army fled when the cannonade began. Aboyne retired on Aberdeen, blocking the only approach to the city— the narrow Bridge of Dee—with turf and stones. The defences were forced and Montrose captured Aberdeen. In 1644, after he had turned Royalist, he was again in the Mearns, marching from his victory at Tippermuir. Crossing the Dee at Mills of Drum, he took Aberdeen. A year later he returned and burned the House of Durris. At Stonehaven he did fearful havoc both by fire and sword, devastating houses, farms, and woods so that " the hart, the hind, the deer, and the roe skirlt at the sicht of the fire," whatever may have been the feelings of the sorely stricken inhabitants. Finding that the Earl and others had secured themselves in Dunnottar Castle, he pillaged and burned the village of Cowie, with the boats and stores, and all the lands of Dunnottar, Fetteresso, Glenbervie, and Arbuthnott. Marching along, he routed a party of the Covenanters at Haulkerton near Laurencekirk, and made the Howe " black with fire and red with blood." His last progress through the Mearns was in 1650, when as a prisoner, bound hand and foot, he was led to his execution in Edinburgh.

After Charles II.'s coronation at Scone, January 1, 1651, the " honours " of Scotland—the crown, the sword, the sceptre—had been deposited in Dunnottar. Dunnottar was the last stronghold to yield to Cromwell's troops. It was invested in the late autumn of 1651. The English general knew that the Regalia had been

E

taken into the castle, while George Ogilvie of Barras, the governor, doubted if he could hold out with his meagre garrison, especially as food was scanty and mutiny was appearing among the men. At this crisis Mrs Granger, wife of the minister of Kinneff, obtained

Regalia
(*Now in Crown Room, Edinburgh Castle*)

permission to visit Mrs Ogilvy. A scheme was devised to save the Regalia. When Mrs Granger left, she had the crown concealed in her lap; and her serving-woman carried the sceptre and the sword in bundles of flax. A touch of irony is added to the incident in the tradition that the English general himself gallantly assisted Mrs Granger to her horse. In a short time the Regalia lay, carefully wrapped up, under the floor of Kinneff church. There they remained till after the Restoration.

In 1685, during the scare of Argyll's invasion, over a hundred Covenanters from the south-west of Scotland

Concealing the Regalia in Kinneff Church

were imprisoned in Dunnottar. Men and women were shut up in a vault too small for them either to lie or sit. It had but one window, and the floor was ankle-deep in mud.   After some time, the men and the women were separated, and several vaults were used instead of one.

Imprisoned Clergyman baptising Children, Stonehaven

From the window of the large vault twenty-five tried to escape down the steep cliff.   Two were killed ;   a few eluded capture ;   those who were unsuccessful were bound and laid on their backs for several hours with burning matches between their fingers. After two months the survivors were conveyed to Leith, where they could choose either to take the Test Act or be banished to the Plantations. Most elected to go into exile.

Like the north-east of Scotland generally, the north of Kincardineshire was strongly Jacobite, due to the influence of the Earl Marischal. In 1715, when the Chevalier de St George—the Old Pretender—was passing south from Peterhead to join his followers, he visited Fetteresso, where he was proclaimed king. In 1746, when the Duke of Cumberland was marching to Aberdeen, ultimately to meet the Young Pretender at Culloden, he burned the Episcopal chapels at Stone-haven, Drumlithie, and Muchalls. The Episcopal clergy, as favouring the Stewarts, were frequently "rabbled" at this time, and some of them imprisoned.

15. Antiquities

The prehistoric period of man's existence is divided by archæologists into the Stone Age, the Bronze Age, and the Iron Age, according to the materials of which implements of industry or weapons of war were constructed. It must not, however, be supposed that bronze implements, when first fashioned, immediately displaced stone implements, or that weapons of iron at once superseded those in previous use. The different periods overlapped, and the introduction of the newer and better implements was gradual.

Of the Old Stone Age no examples have as yet been unearthed in Scotland; but of the Neolithic or New Stone Age examples are everywhere abundant. Axes, arrow-heads, celts, knives of flint, whorls, beads, and buttons of jet are among the ancient treasures found

Bronze Vessels from Banchory Loch

in the county, almost every parish having contributed its quota.

Specimens of the Bronze Age, which began probably about 1200 or 1400 years B.C. and lasted for eight or ten centuries, have also been found, and include spears, hatchets, and other implements. A fine example of a bronze dagger was unearthed in 1840 near the site of the Roman Camp at Fordoun, while similar ones have come from Arbuthnott and Kinneff.

Kincardineshire, especially in the north, has numerous stone circles. Generally the circles consist of huge blocks of stone, irregular and of unequal size, some standing erect, others fallen down, arranged in a circle, which encloses one or even more concentric circles. Sometimes there is in the circle itself, or in the cir-

Auchquhorthies. View from the South

cumference, a large stone, known from the way it lies as the recumbent stone. It is usually on the south side of the circle, and is supposed to have been an altar stone. The circumference of the circles varies a good deal. The diameter of those in Banchory-Devenick is

Auchquhorthies. Recumbent Stone

from 30 to 100 ft., the largest being the well-known circle at Auchquhorthies. This one presents some features of interest. The recumbent stone, 9 ft. 9 ins. long, 5 ft. high, and about 1 foot wide across the top, weighs about 10½ tons. It lies a considerable distance from the standing stones. The stones in the northern arc are small in comparison with the others. The circles are all composed of the blue granite common to the district, and nearly all have their recumbent

Ogham Stone, Auquhollie

stone on the south or south-west, while in more than half the circles relics have been found. What these circles were used for is still a matter of doubt. But since urns and calcined bones have usually been discovered in them, it is likely that they were burial places of the Bronze Age.

Other places of sepulture are the mounds or cairns under which have been found stone cists or coffins containing skeleton remains, along with urns, cups, beads, rings, arrow-heads, and other relics. In Banchory, Strachan, Marykirk, Kinneff, and elsewhere, these have been discovered, bearing mute but expressive testimony to the ideas which prehistoric man had of religion and of a future state.

At Greencairn Castle near Fettercairn are still to be seen traces of what is supposed to have been a vitrified fort. It was oval in form, and, like other strongholds of the same character, was surrounded by two ramparts, built of stone, without any lime or mortar, and without the least mark of any tool, although under the foundation wood ashes were got. Evidence of vitrification of the walls was obtained by Sir Walter Scott in 1796.

Of sculptured stones in the county the most interesting and most ancient is the Ogham stone at Auqu-

Surface of Crannog, Loch of Banchory

hollie, near Stonehaven, one of the fourteen to be found in Scotland. The writing is in some parts much worn and doubtful, but it has been deciphered and translated as follows :

" F[a] dh Donan ui te [? n]
[Here] rests [the body] of Donan, of the race of . . ."

There are three Scottish saints of that name, one being connected with Aberdeenshire.

When, about sixty years ago, the Loch of Banchory, or Leys, in Banchory-Ternan, was drained, an island was found to be artificial—a specimen of the old Celtic lake dwelling or crannog. It rested on trunks of oak and birch trees laid alternately, the spaces being filled up with earth and stones, and the island was surrounded with oak piles to prevent it from being washed away.

There are three interesting examples of old crosses in the county. That at Fettercairn, the old market cross of Kincardine, surmounts an octagonal flight of steps, and has an iron rivet to which criminals in olden days used to be chained by the jougs. The base and shaft of the old cross of Stonehaven stands beside the steeple (itself a picturesque Dutch-like erection dating from 1797). In the square of Bervie is a cross, about 14 ft. high, surrounded by a flight of steps, and supposed to be of considerable antiquity. The county has numerous holy wells, none of them of great importance. Two interesting cup-marked stones are preserved—one at Cowie House, and the other at Dunnottar Manse.

16. Architecture—(a) Ecclesiastical

Though Scotland cannot claim to have originated a new and distinctive style of architecture, yet it can show a continuous series of ecclesiastical buildings, beginning with the simplest and rudest of monkish cells, extending through all the periods of mediaeval art. Of church architecture, however, as we now understand it, there was none during the first seven centuries. It really

Cowie Church

began about the tenth century, when the round towers first appeared.

Of ecclesiastical buildings now in ruins Kincardine-shire has some very interesting examples. Cowie Church, or more correctly the Chapel of St Mary, picturesquely situated a little north of Stonehaven

Arbuthnott Church

Bay, is an example of a simple oblong structure in the first pointed style. There are three fine lancet-pointed windows of the thirteenth century in the east gable, with a square window in the west. The chapel was consecrated in 1276, and was unroofed by ecclesiastical authority shortly before the Reformation on account of scandals.

At the Kirktown of Fetteresso the roofless ruins of the old church of Fetteresso stand upon a knoll

Illumination from Arbuthnott Book of Hours

(St Gregory, when saying mass, has a vision of Our Lord)

which is one of the oldest ecclesiastical sites in the Mearns. The ancient church was dedicated to a Celtic saint of the sixth century named Caran. The pointed doorway on the north side and parts of the adjacent walls belonged to the church which was consecrated by David de Bernham, Bishop of St Andrews, in 1246. Its belfry is a good example of the belfries to be seen in the north-east of Scotland.

The ruined church of St Fittick, about a mile southeast of Aberdeen, stands on the site of an early church which was granted by William the Lyon to his favourite Abbey of Arbroath, and remained attached to it till the Reformation.

The parish church of Arbuthnott, dedicated to St Ternan, is undoubtedly the most interesting piece of ecclesiastical architecture in Kincardineshire. One of the few existing pre-Reformation churches in the north of Scotland, it was consecrated in 1242 by David de Bernham, Bishop of St Andrews. It is long and narrow, consisting of an aisleless nave and chancel, and what is known as the Arbuthnott aisle, which projects from the south side of the chancel. The Arbuthnott aisle—the most striking feature of the exterior—was built in 1505, the west gable of the nave with the circular bell-turret being added at the same time. The aisle is in two stories, the lower a vaulted chapel with an apsidal termination to the south. Within the apse lies a monumental effigy, probably that of James Arbuthnott, who died in 1521. On the side of the base are four shields, bearing the names of Stuart, Arbuthnott, and

St Mary's College, Blairs

Douglas. The chancel is sharply pointed. Three fine stained-glass windows adorn the east gable of the chancel. The church was skilfully restored in 1890, after being accidently burned the previous year. About 1475 the vicar of the parish, James Sibbald, produced three service-books, to which the name of Arbuthnott is attached—a Missal, a Book of Hours, and a Psalter.

Another interesting church, which originally dates from the thirteenth century, is that of Kinneff. The present church, which has suffered from various restorations, has in the east gable a small Norman window, and five Gothic windows in the south wall. The historical interest of the church is even greater than its architectural interest. For it contains several mural monuments, one to Rev. James Granger and his wife Christian Fletcher, who preserved the " honours " of Scotland.

The parish church of Fordoun, a prominent object in the landscape, with its handsome square Gothic tower, nearly· 100 ft. high, was erected in 1830, and is the successor of a very old church, which was demolished in 1787. Beside it is the small chapel of St Palladius, a modern restoration ; but the traditions regarding this saint and his connection with the place as exemplified in chapel, well, and annual fair which bear his name, go back to the fifth century. Within the chapel is a sculptured stone which, according to Professor Stuart, is intended to commemorate the death of Kenneth III.

At Blairs, in Maryculter parish, the Roman Catholic College of St Mary stands conspicuous on a slope overlooking the valley of the Dee. The estate of Blairs was

Blairs Portrait of Mary Queen of Scots

once the property of the Knights Templars. The college possesses a famous portrait of Mary Queen of Scots—an excellent likeness. It may have been painted from a miniature given by Mary on the morning of her execution, to Elizabeth Curle, one of her attendants, who bequeathed miniature and portrait to the Scots College at Douai. In the days of the French Revolution the portrait lay hid in a chimney to save it from the fury of the mob. In the background of the picture, left, there is a sketch of the execution ; and, right, Elizabeth Curle appears.

17. Architecture—(b) Castellated

The architecture of a country is a genuine record of its development and progress in civilisation. In the rude Scotland of early times comfort and convenience were sacrificed for strength and protection from enemies : hence the walls of enormous thickness, the strong gates, the moat, and the ramparts of earth and stone characterising the earliest buildings that survive. Previous to the Norman Conquest, the building of castles with stone and lime was not practised, the earliest fortifications being constructed with earth mounds and wooden palisades on a turf wall. The position of many of the old castles shows that up to the thirteenth century, if not later, the builders of the castles trusted more to water than to hill for their defence. The steep cliff, facing and perhaps projecting into the sea, almost surrounded by the breaking waves, and connected only by a narrow pathway to the mainland, was a typical and well-chosen spot on which to erect a safe resort in time of danger. The Kaim of Mathers near St Cyrus, now a roofless relic of the stronghold of the Barclays, consisted of a tower 40 ft. square and four stories high, perched on the top of a precipitous rock jutting out into the sea. It was built, after the murder of Melville the Sheriff (1420), by the laird of Mathers, who preferred to stay at home, and

> " Buyld a lordlie Kaim,
> All on the stonie rock,
> Which mote defie the sovereign's arms
> Or eke the tempest's shock."

The general appearance of the thirteenth-century castles was that of a huge fortified enclosure. The plan is usually quadrilateral, but more or less irregular to suit the site. Cosmo Innes says that Kincardine Castle, near Fettercairn, was built in the thirteenth century, though it doubtless occupied the site of several previous royal palaces of wood and wattle, where Pictish and Scottish kings held state. The castle was fully 130 ft. square, and had walls of enormous thickness, which were surrounded by marshes across which no enemy could safely venture.

During the fourteenth century, after the Wars of Independence, there was very little castle-building in Scotland. Even had the resources of the country been greater than they were, the nobles were not encouraged by King Robert Bruce to build strong mansions, as those were liable to be captured by the English, and

the King's policy was rather to starve the enemy out
of the country than to fight him. The model of the
castle still remained the square tower or Norman keep
with very thick walls, defended from a parapeted path
round the top of the tower. Gradually the simple
keep was extended by adding on a small wing at one
corner, making the ground plan of the whole building
take the form of the letter L. The entrance was then
placed as a rule at the re-entering angle. The ground
floor was vaulted and used for a store-room. Access
from one story to another was by a narrow corkscrew
stair at one corner in the thick wall. With the outside
entrance raised above the ground level and reached only
by a removable ladder, such towers could resist siege
and fire, and even when taken, could not be easily
damaged.

The tower of Benholm, now a part of Benholm Castle,
is a fine example of the fifteenth-century keep. It is
crowned with a parapet and angle bartisans, and has
on its top a square cape house or watch turret. This is
a primitive indication of the various additions which
were sometimes made on the parapets by raising them
and covering them in with roofs, a feature that may
be seen in several of the later mansions of the Mearns.
Fiddes Castle, formerly a dower-house of the Arbuthnott
family, is a very fine example of the sixteenth-century
castle. The general arrangement is that of the L plan,
but the staircase is projected in a large circular tower
beyond the corner of the main building. Another
circular tower occupies the corresponding angle on the

opposite side of the main building, and a third is corbelled out from the first floor on the north side.

In the latter half of the sixteenth century various influences contributed to a decided change in castellated architecture. With the introduction of artillery the whole idea of defence became altered. Safety from sudden attack was of more consequence than the idea of making the castles impregnable : shelter from the elements was of as great importance as shelter from the enemy. More attention was paid to ornamental detail, and internal comfort and convenience. The grim fortress was gradually transformed into the county mansion, although the keep or quadrangular plan was still adhered to. The change was of course gradual. The castles were built round a courtyard, but turrets were placed at every angle of the building. The lower walls were severely plain. The roofs became high-pitched with picturesque chimneys, dormer windows, and crow-stepped gables. The nobles, enriched by the revenues of the church lands secularised after the Reformation, were enabled either to build new castles or extend the old. The effect of the Union in 1603, after which many of the nobility followed the court to London, was also seen in the higher standard of domestic comfort and house accommodation which imitated that south of the Border. These features are manifest in Crathes Castle, which externally presents a wonderful cluster of pinnacles and turrets at the roof above a plain building with rounded corners below. The corbelling and carving are of a very elaborate and ornate character.

Gargoyles at impossible places, applied as mere ornaments, also occur in profusion. In the east wall over
the doorway, which still presents its original iron " yett,"
are two shields containing the Burnett arms with the

Crathes Castle

dates of the erection and completion of the castle,
1553 and 1596. Balbegno Castle is another interesting
example of a castle on the L plan, into which various
modifications have been introduced. As at Crathes, the
whole of the re-entering angle is filled up, instead of
a turret being inserted in the angle. This is to give
provision for a wide staircase to the first floor. It is
one of the few castles in Scotland which have a ribbed

Part of Vaulted Roof, Balbegno Castle

and groined vault over the hall. The compartments of the vaulting are painted with the armorial bearings of some of the principal families in Scotland.

Muchalls Castle is a well-preserved specimen of the Scottish mansion of the beginning of the seventeenth century. It is designed on a plan of buildings surrounding a courtyard, the north, the east, and part of the west side of the square being occupied with the house, and the remaining side enclosed with a wall. The details of the internal decorations are in the Renaissance style, which began to assert itself in Scotland early in the seventeenth century. The ceiling of the dining-room is the great feature of the house. It is ornamented with ribbed plaster work, the panels being filled in with the heads of Roman emperors, classical heroes, and Scripture characters.

Apart from its romantic and historical associations, Dunnottar Castle is of great architectural interest, for it exhibits the various changes which took place in the disposition of buildings and defences, as well as in the domestic arrangements, from the fifteenth to the seventeenth century. *Dunnottar* signifies a strong promontory, and the situation justifies the name. The castle stands on the platform of an isolated rock washed on three sides by the sea. The perpendicular cliffs rise to a height of 160 ft. except at the narrow strip of land on the level of the seashore, by which it is joined to the mainland. The area of the site is about 3½ acres. A very short steep path leads to the gateway, on the right of which is " Benholm's Lodgings," a five-storied building, furnished with three tiers of loopholes. A strong portcullis had closed down the gateway, within which, to left and right respectively, are the guard-room and the prison, the whole of the open approach being effectively commanded from the buildings above and

Dunnottar from the South

from the parapets. The oldest building is undoubtedly the keep or tower at the south-west corner of the platform, which dates from the early part of the fifteenth century. The plan presents the usual arrangement of the period, the L shape, four stories in height, and with walls 5 ft. thick. Further to the east is an extensive range containing stables and the Priest's House. This part belongs to the latter half of the sixteenth century. The church is next in order. The original church, con-

secrated in 1246, stood on this site. To it the lower
part of the south wall belongs, but the rest of the

Entrance to Dunnottar Castle
(*Benholm's Lodgings on right*)

building was reconstructed early in the sixteenth
century. Part of the church must, therefore, be the
oldest-built work in the castle. The walls were at one
time ornamented with monuments to the Keiths, but

they have all disappeared. The latest addition to the castle is the projecting wing at the north-east corner of the quadrangle. Its ground floor contains a vaulted apartment 58 ft. long and 15 ft. wide. Originally intended as a store-room, it gained an unenviable notoriety as the prison of the Covenanters, or Whigs' Vault. There are curious niches in the walls, apparently intended for the insertion of prisoners' hands when torture was applied for misdemeanours. Below the Whigs' Vault is a smaller one, where, it is said, no fewer than forty-two of the Covenanters were confined for a time. From the Whigs' Vault, by the great staircase, we reach the dining-room, the windows of which give a wide prospect of sea and wild cliffs.

18. Architecture—(c) Domestic

The mansions of the Mearns are not only numerous, but in many respects remarkably elegant structures. Several are of ancient foundation, and have been re-modelled or enlarged as the demands for convenience and domestic comfort grew with the improving spirit of the times. Many of the sites are well-chosen either for beauty or, as in the case of older castles, for defence or observation. Of the castles on or near the coast the most interesting historically is Fetteresso Castle, for-merly the home of the Earls Marischal. It was burned by the Marquis of Montrose in 1645, and rebuilt in 1671 ; but a large part of it is only about one hundred years old. After landing at Peterhead, in 1715, the Chevalier went

Fetteresso Castle

to Fetteresso, where he was warmly received and
hospitably entertained for more than a week. At the
door of the castle he was proclaimed James VIII. by
the Earl Marischal. The castle stands in the Carron
valley near Stonehaven. In the same locality is Urie

Old House of Urie
(Friends' meeting-place on right)

House, a fine mansion in the Elizabethan style, amid
some 700 acres of well-wooded grounds along the Cowie
Water. Urie belonged to the Marischal family, and
then to the Barclays, of Quaker, farming, and pedestrian
fame.

Further south on the coast there is a succession of
mansions. Hallgreen Castle, overlooking Bervie Bay,
dates from the sixteenth century, but has modern

additions. Brotherton Castle, a little to the north of Johnshaven, is a fine building in the baronial style. Lauriston Castle, occupying a picturesque situation in the " Den " of the same name, was built by the Straitons in the thirteenth century. Alexander Straiton, " the knicht of Lauriston," was one of the 500 knights slain at Harlaw in 1411. Ecclesgreig Castle, on a rising ground to the north of St Cyrus village, is modern. Its steep-pitched roof and crow-stepped gables, surmounted by conical turrets, give it a graceful and imposing appearance. The surrounding policies are tastefully laid out.

The Burn House, built in 1791 by Lord Adam Gordon, is romantically situated on the east bank of the North Esk. The house, itself a massive but somewhat plain structure, is surrounded by " woods, walks, and scenes of beauty," as picturesque as any in the county, forming a striking contrast to the " dreary desert " the spot was said to be towards the end of the eighteenth century. Fasque House, a splendid pile built in 1809 in the English baronial style, is a very commodious mansion, and from its elevated situation commands an extensive and magnificent view of the Howe. It is the residence of Sir John Gladstone, the nephew of the late Right Hon. W. E. Gladstone, who as a young man, frequently resided here. Drumtochty Castle, a fine Gothic building, occupies an ideal site on the richly wooded bank of the Luther, opposite Strathfinella Hill. Monboddo House is more interesting historically than architecturally, as the birthplace and residence of Lord Monboddo.

Dr Johnson thus refers to the visit which he paid to Monboddo in 1773 : " Early in the afternoon Mr Boswell observed that we were no great distance from the house of Lord Monboddo. The magnetism of his conversation easily drew us out of our way, and the entertainment which we received would have been a sufficient recompense for a greater deviation." Glenbervie House, on the north side of the Bervie Water, was an ancient seat of the Douglases, and the oldest part dates back to the twelfth century at least. Other mansions in the Howe district are Inglismaldie House in Marykirk, one of the seats of the Earl of Kintore ; Fettercairn House, dating from 1666, but at various times considerably improved and enlarged ; Thornton Castle, also a very old building, about two miles west of Laurencekirk ; Arbuthnott House, on the left bank of the Bervie Water, the seat of the old family of Arbuthnott.

Of the larger mansions on Deeside, Kincausie House, and Durris House are the oldest. The former, beautifully situated on the right bank of the Dee about seven miles from Aberdeen, is surrounded by fine old timber. Durris House, an elegant and very substantially built modern mansion, was erected close to the site of the old castle of Dores, a residence of Alexander III.

19. Communications—Roads and Railways

Kincardineshire being on the direct route between the north and the south of Scotland, the earliest of the

main roads in the county were avenues, running generally north and south, and leading to the Highlands and
Lowlands.   Where much of the land was ill-drained and
boggy, the making of suitable roads was often a difficult
and tedious matter.   The high roads, being the dry
roads, had perforce at first to be followed, while the
straight line as the shortest distance between any two
given points was, where practicable, preferred.   Until
well after the Union of 1707, the roads in Kincardineshire were, as elsewhere in the north of Scotland, in
a very neglected state.   Where wheeled vehicles were
non-existent or few, wide, well-made roads were of little
consequence.   Bridle paths sufficed for the needs of
the pack-horse that plodded along by ways none too
safe by day or night.

The Roman road from Tay to Dee is undoubtedly the
oldest, and its course can be generally traced in the line
of the Roman camps, usually a day's march apart.
Starting probably at Ardoch in Perthshire, and continued through the northern district of Forfarshire, it
entered the county at Kingsford (a modern name) in
a north-easterly direction between the parishes of
Marykirk and Fettercairn ; whence the route was direct
to the camp at the Mains of Fordoun.   From this it was
continued to the camp at Raedykes near Stonehaven,
and thence to Normandykes, Peterculter, where it
crossed the Dee.   At Marykirk a short branch, probably
not, however, a Roman road, struck to the left, leading
to the royal palace of Kincardine.   From that point it
was continued to the pass of Cairn O' Mount, which in

Cairn O' Mount Road

G

later days echoed, not to the tramp of the Roman legions, but to the tread of the red-coated regiments of the second King George, under that renowned road-maker General Wade, the last of whose military roads this was. From the Roman road, or its successors, numerous cross-roads struck off on each side leading to hill and sea. The hill roads were utilised by the Highland drovers on their way to the great annual trysts and fairs south of the Grampians, while the roads that led from the numerous small shipping ports were convenient for transporting either coal or lime into the interior.

For the first three-quarters of the last century the roads were divided into two classes—the turnpike or toll, and the statute labour roads. The former were originally made by subscription, and partly upheld by tolls, while the latter were made and upheld from highway and bridge moneys paid by heritors and others. When the Roads and Bridges Act of 1879 came into force, a road rate was imposed on all householders ; and since then a gradual improvement has been effected on the roads so that they are now, as a rule, very suitable for the needs of modern travelling.

The main road through the county leads from Brechin by North Water Bridge, west of Marykirk, to Laurence-kirk, Fordoun, Stonehaven, and Aberdeen. This is the main route for traffic from Edinburgh, through Strathmore and the Howe of the Mearns. A parallel road to this, but running along the base of the hills, passes through Fettercairn and the beautiful Glen of

Drumtochty, thence through Fordoun, Glenbervie, and Fetteresso parishes to Stonehaven, where it joins the Great North Road. From Montrose a splendid turnpike road runs close to the coast through St Cyrus, Bervie, and Stonehaven, where it also meets the main road. These three parallel roads are connected by numerous cross-roads, which give free access to all parts of the county. One of the best roads in the county is that along the south side of the Dee from Aberdeen to Maryculter, Durris, Banchory, and Strachan. From the coast various cross-roads connect with this road—the well-known " Slug " road from Stonehaven going through Rickarton and Durris to Banchory ; another through Cookney, Netherley, and Maryculter to the Dee valley ; and a third from Portlethen through Fetteresso, Maryculter, Durris, and Strachan.

The county has no canals, though towards the end of the eighteenth century there was much talk of constructing one through the Howe of the Mearns and Strathmore to the Tay. The general opinion on this is pithily summed up by Robertson (*Agricultural Survey*) : " There seems, in fact, to be very little to urge against the practicability of the thing, and nothing perhaps against its expediency, but that it would be of no use. Nobody would think of conveying goods 40 or 50 miles by water who had it in his power to bring them directly to market by an easy land carriage, of less than the fourth part of the distance and time."

The railways in the county run practically parallel and contiguous to the main roads. They belong to three railway companies—the Caledonian, the North British, and the Great North of Scotland. The northern section of the Caledonian, first called the Aberdeen, and afterwards the Scottish North-Eastern, was opened throughout in 1850. It enters the county by a viaduct of thirteen spans over the North Esk near Marykirk Station, and running northward past Laurence-kirk, Fordoun, and Drumlithie, where the highest point on the section is, reaches through heavy cuttings the sea at Stonehaven, after which it follows the coast to Aberdeen. A section of the North British Railway, about 14 miles long, runs from Montrose along the sea to Bervie, at present the terminus, although proposals have been made to connect it with Stonehaven by a light railway. From Kinnaber Junction, two miles north of Montrose, where the North British and Caledonian main lines connect, the former company possesses certain running powers over the Caledonian system to Aberdeen. The Deeside railway, owned by the Great North of Scotland Company, runs from Aberdeen along the north side of the Dee. It enters the county near Crathes Station, 14 miles from Aberdeen, and leaves it close to Glassel Station.

20. Administration and Divisions

Sheriffs were appointed in the twelfth century, but it was not till the fourteenth that the office became hereditary in Scotland. In Kincardineshire the Keiths were the hereditary sheriffs for some two hundred years from about 1350. Their jurisdiction probably did not coincide with the present boundaries, but their power and influence in the county was undoubted. It would appear also that in the Mearns the offices of sheriff and forester were often united. The royal forester had jurisdiction in offences against the forest laws, and received certain payments or privileges for superintending the hunting domains, such as Cowie and Durris. In addition to the sheriff, we hear also of thanes, of whom there were at least seven in the Mearns. Originally stewards over the royal lands, they ultimately became hereditary tenants of the King. Those hereditary powers were abolished after the '' Forty-five,'' the sheriff, an advocate by profession, henceforth holding his office direct from the Crown.

Besides the Lord-Lieutenant, who may be regarded as the head of the county, but whose duties are now largely ceremonial, there are in Kincardineshire Deputy-Lieutenants ; but the real executive power is vested in the salaried Sheriff, assisted in his judicial and administrative capacity by a Sheriff-Substitute. The Sheriff-Principal of Aberdeenshire is Sheriff of Kincardine and also of Banff.

The chief administrative body in the county is the County Council, which came into existence in 1889. It is presided over by a chairman chosen from amongst the elected members, who is also designated convener of the county. Representatives come from each of the nineteen parishes or electoral divisions in the county, these again being grouped into five districts :

(1) Laurencekirk district, with four electoral divisions ;
(2) St Cyrus district, with three ; (3) Stonehaven district, with five ; (4) Lower Deeside, with four ; and
(5) Upper Deeside, with three. Each of the five districts
has a committee consisting of the County Councillors for
the electoral divisions of the district and of representatives selected from the various parish councils. Roads
and bridges, public health, diseases of animals, protection of wild birds, valuation, finance, and the general
administrative oversight of the county are under the
control of the County Council.

By the Education Act of 1872, School Boards in every
parish had the charge of education ; but the Education
Act of 1918 has now established an Education Authority
for the whole county to control both primary and
secondary schools.

The civil parishes, each with its council to carry out
the provisions of the Poor Law and other duties, number
nineteen : Arbuthnott, Banchory-Devenick, Banchory-
Ternan, Benholm, Bervie, Dunnottar, Durris, Fetter-
cairn, Fetteresso, Fordoun, Garvock, Glenbervie, Kinneff,
Laurencekirk, Maryculter, Marykirk, Nigg, St Cyrus,
Strachan. The ecclesiastical parishes are twenty-two :
all the civil parishes and the *quoad sacra* parishes of
Cookney, Portlethen, and Rickarton. Fifteen of these
form the Presbytery of Fordoun, while five are in the
Presbytery of Aberdeen and two in the Presbytery of
Kincardine O' Neil.

The county now unites with the Western Division of
Aberdeenshire in returning one member to Parliament.

Bervie, a very ancient burgh, sent representatives to the Scottish Parliament from 1612, at least, down to 1707. Under the Act of Union it was classed with Aberdeen, Arbroath, Brechin, and Montrose—a group returning one member to the British Parliament. Bervie is still one of the Montrose Burghs, Aberde n has two members of its own.

21. Roll of Honour

Though small in size, Kincardineshire has a remarkable muster-roll of notables whose reputation is by no means local.

To begin, it claims to be the cradle of the family to which Robert Burns belonged. For many generations Glenbervie had been the home of the family of Burness, as the name was invariably spelled; and it was from Clochnahill, near Stonehaven, that the poet's father set out to better his fortunes in the south. It was from his father that Robert Burns inherited his brain-power, his hypochondria, and his general superiority. Robert's cousin, John Burness (1771–1826), was author of *Thrummy Cap*, which Burns thought " the best ghost story in the language."

Sir Walter Scott's connection with the county, though not so close or direct, is nevertheless interesting. Readers of the Waverley Novels will remember that it was in the churchyard of Dunnottar that Scott in 1796 first saw Robert Paterson, the original of *Old Mortality*, " engaged in his daily task of cleaning and repairing the ornaments and epitaphs upon the tomb " of the Covenanters. Scott, however, became the begetter of one of the best-known men of the Mearns, the renowned mercenary soldier Captain Dugald Dalgetty, whose " natural hereditament of Drumthwacket " was " the long waste moor so called, that lies five miles south of

Burying-place of Burns's Ancestors

Aberdeen," and who was naturally an alumnus of Marischal College.

Whether as soldiers, administrators, courtiers, or patriots, various members of the Keith family wielded great influence, not only in the county but also throughout the country, from the eleventh century to 1718, when the last Earl Marischal's estates were forfeited to the Crown. This Earl's younger brother, James Keith, after military service with the Spaniards and the

Russians, went to Prussia, where Frederick the Great
at once made him field-marshal, and relied greatly on

George Keith, Fifth Earl Marischal
(*Founder*, 1593, *of Marischal College, Aberdeen*)

his military genius. In 1758 Keith was killed at Hoch-
kirch while for the third time charging the Austrians.

The Falconers, whose name came from the Crown
office they held, were connected with the county in the
twelfth century. Three of them became senators of
the College of Justice, or Lords of Session. One of these

was deprived of his seat, 1649–1660, for " malignancy,"
which drew from Drummond of Hawthornden a sonnet

Field-Marshal James Keith
(*From a painting in the Burgh Council Chamber, Stonehaven*)

in praise of his character and a lament for his misfortunes.
One would like to head the list of historians with the
name of John of. Fordun, author of the important
Scotichronicon; but that he was born in the parish of

Fordoun is merely an inference from his name. He flourished in the fourteenth century. Cosmo Innes (1798–1874), a native of Durris, was trained as a lawyer. In 1846 he was appointed Professor of Constitutional Law and History in Edinburgh University. He is best known for his two historical works—*Scotland in the Middle Ages*, and *Sketches of Early Scotch History*. Dr Cramond, a voluminous writer of histories dealing chiefly with the north-east of Scotland, belonged to Fettercairn.

James Burnett (1714–1799), Lord Monboddo, was famous not merely as a lawyer but also as a litterateur. He first came into prominence as counsel for the Douglases in the Douglas case, and in 1767 he was made a Lord of Session, a position he held for thirty years. His *Origin and Progress of Language*, in which he anticipated the Darwinian theory, is very learned and acute, but very eccentric. Lord Neaves, a versatile successor in the Court of Session, sings of him :

> " His views, when forth at first they came,
> Appeared a little odd O !
> But now we've notions much the same,
> We're back to old Monboddo.

> " Though Darwin now proclaims the law,
> And spreads it far abroad O !
> The man that first the secret saw,
> Was honest old Monboddo."

Lord Gardenstone, another Lord of Session, was like Monboddo, somewhat eccentric, but did much for the village of Laurencekirk, which he got erected into a

Burgh of Barony. Still another judge was Sir John Wishart, who died in 1576, a native of Fordoun. He

James Burnett, Lord Monboddo

was a comrade of Erskine of Dun in the days of the Reformation, and fought at Corrichie.

Of ecclesiastical dignitaries the county can show a

generous muster-roll, an outstanding feature being the relatively large number of bishops. One was Bishop Wishart of St Andrews. Bishop Mitchell, a native of

Dr Thomas Reid

Garvock, was deprived of his office in 1638, and during his exile in Holland worked as a clockmaker. Bishop Keith (1681–1757) was born at Uras, and held the See of Fife. He compiled a valuable history of Scottish affairs from the beginnings of the Reformation to Mary's departure for England in 1568. Gilbert Burnett, Bishop of Salisbury and friend of William III., was a descendant of the Burnetts of Crathes. Alexander Arbuthnott (1538–1583), son of Andrew Arbuthnott of Pitcarles, became Principal of King's College, Aberdeen, in 1569, and soon after received the living of Arbuthnott. Dr James Sibbald, who died about 1650, was a Mearns man. He was minister of St Nicholas, Aberdeen, and a stout opponent of the Covenant. Equally stout on the other side was Rev. Andrew Cant (1590–1663), a native of Strachan. Another native of Strachan was Dr Thomas Reid (1710–1796), parish minister of New Machar in Aberdeenshire and Professor of Philosophy at King's College, Aberdeen. He wrote a renowned book— *Inquiry into the Human Mind on the Principles of Common Sense*—and created the Scottish school of philosophy opposed to David Hume. He succeeded Adam Smith in Glasgow.

In literature the greatest name is Dr John Arbuthnot (1667–1735), son of an Episcopalian clergyman at Arbuthnott. One of the Queen Anne wits and the friend of Swift and Pope, he wrote the *History of John Bull* and was the chief author of the *Memoirs of Martinus Scriblerus.* " The Doctor," said Swift, " has more wit than we all have, and his humanity is equal to his wit." Dr James Beattie (1735–1803), a native of Laurencekirk, and schoolmaster of Fordoun, was appointed to the Chair of Moral Philosophy in Marischal College, Aberdeen. His *Essay on Truth* had a great reputation, while his Spenserian poem *The Minstrel* still finds readers. George Beattie (1786–1823), author of *John of Arnha*,

was a native of St Cyrus. Thomas Ruddiman (1674–
1757), for five years schoolmaster of Laurencekirk, was
a famed Latinist, whose *Rudiments* had great vogue

Dr John Arbuthnot

for many years. David Herd (1732–1810), who belonged
to Marykirk and edited the first classical collection of
Scottish Songs ; Dean Ramsay (1793–1872), author of
Reminiscences of Scottish Life and Character ; Dr John
Longmuir (1803–1883), historian of Dunnottar Castle ;
and Dr John Brebner (1833–1902), a native of Fordoun,

organiser, and for twenty-five years head, of the educational system in the Orange Free State, cannot be left unnamed.

Captain Robert Barclay
(On his walk of a thousand miles)

Various members of that family of strong men, the Barclays of Urie, achieved fame in different ways. The first was Colonel David Barclay, an old soldier of

Gustavus Adolphus, who purchased Urie.   He turned
Quaker, and was accordingly persecuted.   Readers of
Whittier will remember the poem beginning :

> " Up the streets of Aberdeen,
>   By the kirk and college-green,
>   Rode the Laird of Urie."

Dying in 1686, he was succeeded by his son Robert
(1648–1690), who in 1672 had walked in sackcloth through
Aberdeen as a protest against the wickedness of the
times.   Robert was an eminent man, and his *Apology*
is the standard exposition of the principles of the Friends.
A descendant of his, who died in 1790, was the famous
agriculturist ; while another, Captain Robert Barclay
(1779–1854), was a noted pedestrian, whose feat of
walking 1000 miles in 1000 consecutive hours took place
at Newmarket in 1809.

22. The Chief Towns and Villages of Kincardineshire

(The figures in brackets after each name give the population
in 1911, and those at the end of each section are
references to pages in the text.)

Auchinblae, a picturesquely situated village 2 miles north
of Fordoun Station, is a famous summer resort.   Here is
the entrance to the beautiful Glen of Drumtochty.   (p. 56.)

Banchory (1633), in the parish of Banchory-Ternan, was
founded in 1805 and is now a Police Burgh.   The most
popular of resorts on Lower Deeside, it is pleasantly situated
on the north bank of the Dee, 18 miles west of Aberdeen.

H

The picturesque Falls of Feugh are less than a mile from the town. The Nordrach-on-Dee Sanatorium stands in pine woods a little to the west. The Hill of Fare, to the north, was the scene of the battle of Corrichie. (pp. 5, 7, 10, 14, 22, 47, 55, 72, 74, 99.)

Bervie (1173), formerly and still officially Inverbervie, a royal burgh since 1362, has prosperous flax-spinning mills.

Nordrach-on-Dee Sanatorium, Banchory

Salmon - fishing is successfully carried on. David II. landed here in 1341, after his exile in France. (pp. 5, 7, 12, 18, 55, 57, 74, 99, 100, 103.)

Catterline is a fishing hamlet in Kinneff parish, midway between Stonehaven and Bervie. Todhead lighthouse is near. (pp. 40, 58.)

Cove, a fishing village about 4 miles south of Aberdeen, has also fish-manure works. (pp. 36, 56, 58, 101.)

Cowie, a fishing hamlet 1 mile north of Stonehaven. (pp. 20, 23, 37, 58, 65, 76, 101.)

Drumlithie (207), an irregularly built village in Glenbervie parish. The steeple, erected in 1777, is a circular tower surmounted by a belfry. Drumlithie became a Burgh of Barony in 1329. (pp. 69, 100.)

Fettercairn, in the centre of a good agricultural district, is a Burgh of Barony, 5 miles north of Laurencekirk. It has a Gothic arch erected to commemorate the visit of

Mending Nets, Gourdon

Queen Victoria and Prince Albert in 1861; and also a turreted fountain tower, a memorial to Sir John Hepburn Stewart Forbes, Bart. (1804–1866). The old market cross of Kincardine stands in the village. (pp. 13, 47, 48, 55, 56, 61, 72, 74, 83, 98.)

Findon, a village between Cove and Portlethen, the original home of the well-known "Finnan haddock." (p. 36.)

Fordoun, a village with station on the Caledonian Railway line. The parish has historical associations with St Palla-

H*

dius, Lord Monboddo, and James Beattie the poet ; and contains the site of the old county town, Kincardine. The chief village is Auchinblae. (pp. 1, 13, 57, 61, 63, 70, 80, 98, 100, 108, 110, 111.)

Gourdon, a fishing village 1 mile south of Bervie, has a flax mill. (pp. 5, 40, 55, 58, 59.)

Johnshaven, a fishing village and coastguard station in the parish of Benholm, has also a spinning mill. (pp. 5, 40, 55, 58, 59, 94.)

Laurencekirk (1438), a Burgh of Barony, has a large local country trade, a flourishing weekly mart, a brewery, coach works, and some handloom weaving. The renowned Latinist, Thomas Ruddiman, was for a few years schoolmaster here. (pp. 11, 13, 52, 55, 56, 65, 98, 100, 107, 110, 111.)

Luthermuir, a small village in Marykirk, dating from 1771, had formerly handloom weaving.

Marykirk, a village beautifully situated on the left bank of the North Esk, a short mile from Craigo railway station. (pp. 13, 17, 72, 95, 96, 100, 111.)

Muchalls, a neat little village and coastguard station 4 miles north of Stonehaven, is famed for its rock scenery and is much frequented by summer visitors. (pp. 13, 29, 30, 36, 69, 88.)

Portlethen, a small fishing village, 6 miles south of Aberdeen. (pp. 30, 36, 53, 58, 99.)

St Cyrus, a village with a salmon-fishing station in the S.E. corner of the county, was formerly called Ecclesgreig. Both *St Cyrus* and *Ecclesgreig* contain the name of a king of the Scots towards the close of the ninth century, Grig or Girig, who won the title of " Liberator of the Scottish Church." (pp. 12, 21, 24, 28, 40, 52, 57, 63, 83, 94, 99, 111.)

Stonehaven Harbour

Skateraw, a small fishing village, close to Newtonhill railway station. (pp. 36, 58.)

Stonehaven (4266), stands on the bay some 14 miles S.S.W. of Aberdeen, at the mouths of the Carron and the Cowie. In the beginning of the seventeenth century it superseded Kincardine as the county town, and in 1889 was made a Police Burgh. It consists of an old and a new town. The old town, south of the Carron, in Dunnottar parish, is irregularly built, and inhabited mostly by fishermen. The new town, in Fetteresso parish, lies between the two streams. It is regularly laid out and well built. Prominent in the central square is the market house with its lofty steeple, and in Allardyce Street the Italianate town hall. Other notable buildings are the two parish churches and the other churches—United Free, Episcopalian, and Roman Catholic. The Mackie Academy was opened as a Secondary School in 1893. The fishing industry is important ; but for general trade the harbour admits only small vessels. Of recent years Stonehaven has been much resorted to by summer visitors, attracted, for health and pleasure, by its bracing climate, fine cliffs and woods, sea-bathing and boating, golf course and recreation ground. (pp. 5, 7, 19, 21, 22, 37, 44, 45, 47, 49, 55, 56, 58, 60, 61, 64, 65, 69, 73, 74, 93, 96, 98, 99, 100, 102, 103.)

Strachan, or Kirkton of Strachan, a village 4 miles from Banchory-Ternan, is in the largest and hilliest parish. At the western boundary of the parish is Mount Battock, the converging point of three shires—Kincardine, Forfar, Aberdeen. Famous natives were Rev. Andrew Cant and Dr Thomas Reid. (pp. 4, 72, 99, 110.)

Torry (11,428), which less than fifty years ago was a small fishing village, is now an important ward of Aberdeen. It unites with the city for parliamentary, municipal, and educational purposes. The construction of the Victoria Bridge, to take the place of the ferry, and the introduction of trawl-fishing led to the rapid growth of Torry. (p. 50.)

DIAGRAMS 119

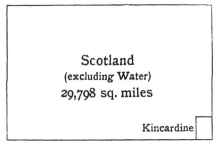

Fig. 1. Area of Kincardineshire (382 square miles)
compared with that of Scotland

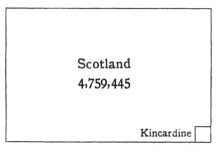

Fig. 2. Population of Kincardineshire (41,007) compared
with that of Scotland at the last Census

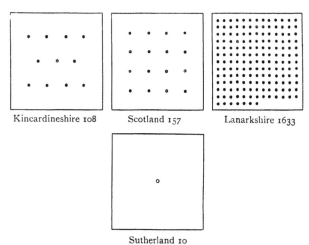

Kincardineshire 108   Scotland 157   Lanarkshire 1633

Sutherland 10

Fig. 3. Comparative density of Population to the square
mile at the last Census

(*Each dot represents ten persons*)

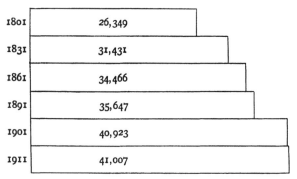

1801	26,349
1831	31,431
1861	34,466
1891	35,647
1901	40,923
1911	41,007

Fig. 4. Growth of Population in Kincardineshire

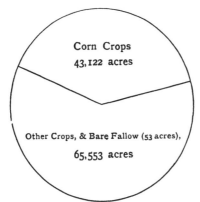

Fig. 5. Proportionate area under Corn Crops compared with that of other cultivated land in Kincardineshire

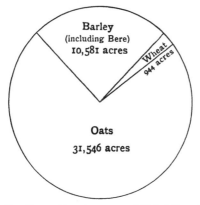

Fig. 6. Proportionate areas of Chief Cereals in Kincardineshire

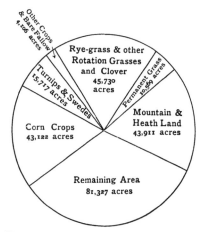

Fig. 7.   Proportionate areas of Land in
Kincardineshire

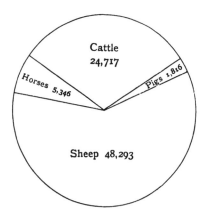

Fig. 8.   Proportionate numbers of Live Stock
in Kincardineshire

www.ingramcontent.com/pod-product-compliance
Ingram Content Group UK Ltd.
Pitfield, Milton Keynes, MK11 3LW, UK
UKHW042146280225
455719UK00001B/143

9 781107 649705